Placental Proteins

Edited by
A. Klopper and T. Chard

With 65 Figures and 36 Tables

Springer-Verlag
Berlin Heidelberg New York 1979

Arnold Klopper
Professor of Reproductive Endocrinology
Department of Obstetrics and Gynaecology, University of Aberdeen, Royal
Infirmary, Foresterhill, Aberdeen AB9 2ZD (U.K.)

Timothy Chard
Professor of Reproductive Physiology
Joint Academic Unit of Obstetrics, Gynaecology, and Reproductive Physiolo-
gy, The Medical College of St. Bartholomew's Hospital, West Smithfield,
London EC1A 7BE (U.K.)

ISBN-13: 978-3-540-09406-7   e-ISBN-13: 978-1-4471-1290-7
DOI: 10.1007/ 978-1-4471-1290-7

Library of Congress Cataloging in Publication Data. Main entry under title: Placental
proteins. Bibliography: p. Includes index. 1. Placenta. 2. Proteins. 3. Pregnancy.
I. Klopper, Arnold. II. Chard, T. QP281.P55 612.6'3 79-13071

2128/3140-543210

Preface

In July 1978 a group met in Aberdeen to discuss the whole range
of new proteins recently isolated from the human placenta. With
the exception of Yuri Tatarinov all the main pioneers in the
field were present, and this book arose from the discussions
which took place. Each author was asked to bring a written man-
uscript corresponding to but not necessarily identical with
their verbal presentation.

Nobody was given a specified remit, for the reason that the
subject is so new that it would be impossible to design the
meeting or the book in advance. Each speaker was left free to
put on display whatever he thought was interesting or important
about the newly isolated proteins. Inevitably this has led to
much overlap, since everybody tends to follow the same path at
first. Nevertheless, we shall probably never achieve so much
agreement again. Only Vernon Stevens was set a fixed title outside the immediate field of new placental proteins. This arose
from the very exciting possibility that the new proteins could
be used to induce an autoimmune state to products of conception
and thus serve as the basis for a new method of contraception.
There are few findings at present which bear specifically on
this proposal, but the experience of Vernon Stevens with hCG
could serve as a model of the problems that might be encountered
with SP_1 and PAPP-A.

A brief glance at this book suffices to show the immense problems of nomenclature. Universally acceptable names for any of
the new proteins are unlikely until they can be given a functional designation, and it may be many years before knowledge
is sufficient to permit this. Designation by chemical characteristics also presents a dilemma. It is tempting to refer to
Schwangerschafts protein (SP_1) as pregnancy-specific β_1 glycoprotein ($PS\beta_1G$). But where does this leave pregnancy-associated
plasma protein B, (PAPP-B), which is also a pregnancy-specific
glycoprotein with β electrophoretic mobility? Extensive discussion
at the meeting about nomenclature was inconclusive, and for the
moment SP_1 must top the list of possible names for this protein.

The situation with standards is no better. Clearly the optimum
would be a weighed amount of pure protein and it is to be hoped
that sooner or later some international body such as WHO would
produce such a standard for each new placental protein. In the
interim neither the expedient of an artificially constructed

serum pool nor the device of local laboratory serum pools is completely satisfactory. It was decided that a pool of late pregnancy serum should be collected in Aberdeen and that this should constitute a temporary but arbitrary common unit for all the new proteins. This has been done and the standard is available to all researchers.

It is likely that there will be an explosion of interest in the new placental proteins in the next few years. Clinicians will want to know whether their concentration in maternal blood bears any relevance to placental function and there is already a spate of publications on this topic. This is perhaps unfortunate in the absence of a clear definition of what aspect of placental function is measured. The truly exciting question about the new placental proteins is: what do they do? or for that matter, do they do anything?

April 1979   Arnold Klopper, Aberdeen
                                    Tim Chard, London

Nomenclature of New Pregnancy-"Specific" Plasma Proteins

In the past, considerable confusion has sometimes arisen in the literature because a single protein has been investigated under a variety of names. In order to avoid this difficulty with the new pregnancy-"specific" plasma proteins, a discussion was held at the symposium regarding terminology, so that this problem could be avoided in the near future. The possibility of establishing a new and systematic nomenclature sequence was briefly explored, but it was the group's consensus of opinion that such a change would only increase the present difficulties. It was therefore decided that the current terminology most widely used should be retained as the preferred name, until such time as a clear function for each is established. At that time, each should be renamed to describe its function in pregnancy.

The proteins specifically discussed are the following, with their preferred designations underlined (ref.).

Preferred designation	Alternative names used
1. SP$_1$ (Schwangerschaft spezifisches protein-1)[1]	Trophoblast-specific β_1 globulin[2] PAPP-C (pregnancy-associated plasma protein-C)[3] PSβG (pregnancy-specific β_1 glyco-protein)[4] β_1SP$_1$[5]
2. PAPP-A (pregnancy-associated plasma protein-A)[3]	
3. PAPP-B (pregnancy-associated plasma protein-B)[3]	
4. PP-5 (placental protein-5)[6]	

1 Bohn, H.: Arch. Gynaekol. 210, 440 (1971).
2 Tatarinov, Y.S., Sokolov, A.V.: Int.J.Cancer 19, 161 (1977).
3 Lin, T.M., Halbert, S.P., Kiefer, D., Spellacy, W.N., Gall, S.: Am. J. Obstet. Gynecol. 118, 223 (1974).
4 Towler, C.M., Horne, C.H.W., Jandial, V., Campbell, D.M., MacGillivray, T.: Br. J. Obstet. Gynaecol. 83, 775 (1976).
5 Searle, F., Leake, B.A., Bagshawe, K.D., Dent, J.: Lancet 1978 i, 579.
6 Bohn, H., Winckler, W.: Arch. Gynaekol. 223, 179 (1977).

Contents

List of Contributors

Bischof, P.: Departments of Obstetrics and Gynaecology and of Biochemistry, University of Aberdeen, University Medical Buildings, Foresterhill, Aberdeen AB9 2ZD (U.K.)

Bohn, H.: Research Laboratories of Behringwerke AG, D-3550 Marburg (FRG)

Bremner, R.D.: Departments of Pathology and of Obstetrics, University of Aberdeen, University Medical Buildings, Foresterhill, Aberdeen AB9 2ZD (U.K.)

Chard, T.: Joint Academic Unit of Obstetrics, Gynaecology, and Reproductive Physiology, The Medical College of St. Bartholomew's Hospital, West Smithfield, London EC1A 7BE (U.K.)

Davidson, I.: Department of Obstetrics and Gynaecology, University of Aberdeen, University Medical Buildings, Foresterhill, Aberdeen AB9 2ZD (U.K.)

Glover, R.G.: Department of Pathology and of Obstetrics, University of Aberdeen, University Medical Buildings, Foresterhill, Aberdeen AB9 2ZD (U.K.)

Gordon, Y.B.: Department of Obstetrics and Gynaecology, Royal Free Hospital, Pond Street, Hamstead, London NW3 2QG (U.K.)

Grudzinskas, J.G.: Departments of Obstetrics and Gynaecology, and of Reproductive Physiology, The Medical College of St. Bartholomew's Hospital, West Smithfield, London EC1A 7BE (U.K.)

Halbert, S.P.: Department of Pediatrics, University of Miami, School of Medicine, P.O. Box 875, Biscayne Annex, Miami, FL 33152 (USA)

Horne, C.H.W.: Departments of Pathology and of Obstetrics, University of Aberdeen, University Medical Buildings, Foresterhill, Aberdeen AB9 2ZD (U.K.)

Jandial, V.: Departments of Pathology and of Obstetrics, University of Aberdeen, University Medical Buildings, Foresterhill, Aberdeen AB9 2ZD (U.K.)

Klopper, A.: Department of Obstetrics and Gynaecology, University of Aberdeen, Royal Infirmary, Foresterhill, Aberdeen AB9 2ZD (U.K.)

Kukulska, B.M.: Institute of Genetics, University of Glasgow, Church Street, Glasgow G11 5JS (U.K.)

Lenton, E.A.: University Department of Obstetrics and Gynaecology, Jessop Hospital for Women, Leavygreave Road, Sheffield S3 7RE (U.K.)

Lin, T.-M.: Department of Pediatrics, University of Miami, School of Medicine, P.O. Box 875, Biscayne Annex, Miami, FL 33152 (USA)

Nicholson, L.V.B.: Institute of Genetics, University of Glasgow, Church Street, Glasgow G11 5JS (U.K.)

Obiekwe, B.C.: Departments of Obstetrics and Gynaecology, and of Reproductive Physiology, The Medical College of St. Bartholomew's Hospital, West Smithfield, London EC1A 7BE (U.K.)

Paterson, W.F.: Institute of Genetics, University of Glasgow, Church Street, Glasgow G11 5JS (U.K.)

Smith, R.: Department of Obstetrics and Gynaecology, University of Aberdeen, University Medical Buildings, Foresterhill, Aberdeen AB9 2ZD (U.K.)

Stevens, V.C.: Department of Obstetrics and Gynecology, 410, 10th Avenue, Columbus, OH 43205 (USA)

Sutcliffe, R.: Institute of Genetics, University of Glasgow, Church Street, Glasgow G11 5JS (U.K.)

Tatarinov, Y.S.: Department of Biochemistry and Immunochemical Laboratory for Research on Malignant and Embryonal Tissues, Second Moscow Medical Institute, Moscow G-435 (USSR)

Tatra, G.: Universitäts-Frauenklinik, Spitalgasse 23, A-1090 Vienna (Austria)

Towler, C.M.: Departments of Pathology and of Obstetrics, University of Aberdeen, University Medical Buildings, Foresterhill, Aberdeen AB9 2ZD (U.K.)

1 The Specific Proteins of the Human Placenta Some New Hypotheses

Y.B.Gordon and T.Chard

The human placenta produces a wide range of proteins and small peptides (Table 1.1) some of which are "specific" in the sense of having no obvious identical counterpart in the adult, while others, particularly the small peptides, appear to be identical to adult products. This range of materials is of immense biological and clinical interest: first, because they reflect the function of an organ which for many months is the only lifeline to the fetus; second because their measurement is increasingly widely used as an index of the adequacy of this lifeline and third, because the placenta and its products provide so obvious a model for most cancers.

The present symposium will touch on all these aspects of work on the placental proteins in some detail. However, in this opening section we wish to put forward a set of generalisations about the specific placental proteins. Several of these generalisations may prove to be highly contentious, yet we present them as a basis for discussion and, where existing evidence is not adequate for proof or disproof, as a possible basis for further experiment.

1.1 The Placenta Is an Independent Organism

The philosophy on which this chapter is based is that the placenta is an independent organism geared to maintaining itself and independent of feedback systems from the mother or the fetus (SILMAN, 1978). The cells which give rise to the placenta have differentiated prior to implantation and the morphological differences between the inner cell mass, which forms the embryo, and the trophoblast which forms the placenta, are obvious at the blastocyst stage (EDWARDS, 1977). The placenta will form and secrete specific proteins in the absence of a fetus in anembryonic pregnancy (BENNETT et al., 1978). The maternal uterus provides the implantation site and the blood supply essential to survival of the placenta but the control mechanisms are probably inherent and feedback from the mother has never been convincingly demonstrated.

1.2 All Placental Proteins Are Produced by the Syncytiotrophoblast

This observation is not usually thought to be contentious. Yet the only experimental evidence which supports it comes from immunohistochemical studies (CURRIE et al., 1966; HORNE et al., 1976; LIN and HALBERT, 1976; SCIARRA et al., 1963), the basis of which has been questioned because of the lack of appropriate controls and because a theoretical analysis suggests that the actual concentrations in the trophoblast cytoplasm would not be accessible to current techniques (GAU and CHARD, 1976). All indirect evidence would appear to support the hypothesis of a trophoblastic origin. For reasons detailed in a later section, fetal synthesis is most unlikely and this would include all tissues internal to the basement membrane of the trophoblast. Maternal components of the placenta (decidua) might be a potential source but there is no evidence for this, nor grounds for supposing that an adult non-neoplastic tissue would develop such a wide and specific range of synthetic capabilities. Finally, it has never been questioned that the trophoblast possesses the full machinery for protein synthesis and does so in great abundance (Fig.1.1).

The architecture of the chorionic villus is a key factor in the control of placental protein synthesis and transport. The syncytiotrophoblast is in direct contact with the intervillous blood and is separated from the fetal capillaries by a basement membrane. The proximity of maternal blood to the trophoblast probably acts as the stimulus to the rate of production and release of the protein molecules.

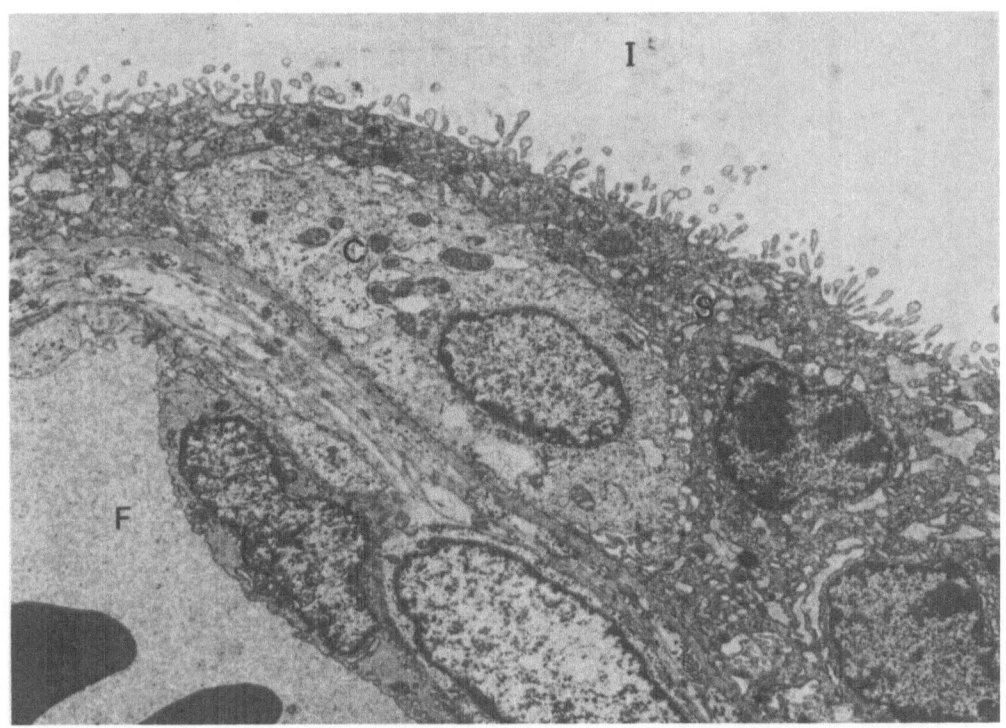

Fig. 1.1. An electron micrograph of human term placenta showing the micro-
villous surface of the syncytiotrophoblast (S) which extends into the
intervillous space (I) the rough endoplasmic reticulum, site of protein
synthesis, is well developed. The cytoplasm of the cytotrophoblast (C)
shows fewer organelles. A fetal capillary (F) lies to the left (courtesy
of Dr.G.Gau, Queen Charlotte's Hospital, London)

The location of the protein synthesising mechanism within
the trophoblast has been the subject of a number of studies.
Prominent among these are the observations of BURGOS and
RODRIGUES (1966) of a clear differentiation between "thick" areas
which are rich in endoplasmic reticulum and microvilli and may
therefore be specialized in protein synthesis, and "thin" areas
without microvilli, known as vasculosyncytial membranes (GETZOWA
and SADOWSKY, 1950), which overlie fetal capillaries and appear
to specialize in the transport of nutrients and waste products
(Fig. 1.2). This dissociation of activities could be of considerable clinical significance since it is possible to visualize
a pathological process which might damage the all-important
"transfer" areas while leaving the "synthetic" areas intact; in
this event a functional insufficiency of the placenta would not
be apparent from biochemical observations of placental synthesis.
In reality, however, trophoblast pathology would seem to be
largely secondary to other factors, principally the occlusion of
maternal decidual arterioles or fetal capillaries in the chorionic
villus; either might be expected to affect all areas of the
trophoblast in equal measure.

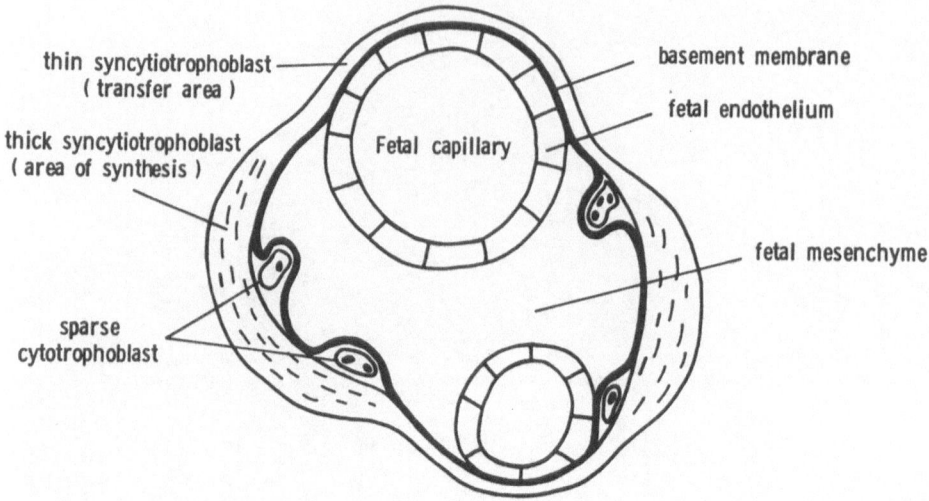

maternal intervillous
space

thin syncytiotrophoblast
(transfer area)

thick syncytiotrophoblast
(area of synthesis)

Fetal capillary

basement membrane

fetal endothelium

fetal mesenchyme

sparse
cytotrophoblast

Fig. 1.2. A diagrammatic representation of the syncytiotrophoblast and fetal
capillary showing a thick area specialised for protein synthesis and a thin
area (vasculo-syncytial membrane) specialised for transfer

An important, but little considered factor in the secretion of
placental proteins into the maternal circulation is the total
area of the trophoblast surface membrane available for secretion.
This membrane is normally characterised by "turf" microvilli
(HAMILTON and HAMILTON, 1977), which must massively increase the
effective area of transfer. Conversion of the "turf" to a relatively smooth surface, which is a feature of some pathological
processes (FOX, 1967) would much reduce the surface area and
hence the potential for secretion.

1.3 Placental Proteins Are Secreted Exclusively into the Mother

Without exception, the levels of the specific placental proteins
in the fetal circulation are 100- to 1000-fold less than the
levels in the maternal circulation. The differential becomes
even more striking when it is appreciated that equivalent secretion into the two circulations would yield higher levels in
the fetus because of its smaller volume.

The most likely reason for this differential is the existence of
a barrier between the trophoblast and the fetal blood, consisting
of basement membrane and capillary endothelium, whereas no
barrier exists on the maternal side (Fig. 1.3). This presupposes
that the placental proteins could be produced and secreted

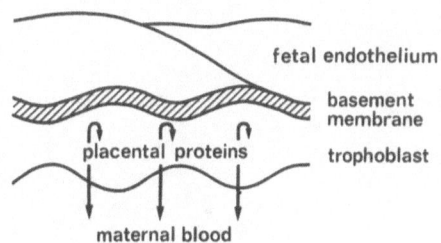

Fig. 1.3. The basement membrane constitutes a barrier to the entry of placental proteins into the fetal circulation.
There is no such barrier to maternal
blood and placental proteins are secreted exclusively into the mother

equally well from any part of the trophoblast cytoplasm, though
it is not impossible that there is specific transport in the
direction of the maternal surface. Specific protein transport
by means of pinocytosis is well described for immunoglobulins
(GITLIN and GITLIN, 1973); however, these molecules are moving
in the opposite direction to placental proteins, towards the
fetus.

If the simple barrier hypothesis is correct then it would be
expected that disruption of the barrier could lead to elevated
levels of the specific protein in the fetal circulation, a situation similar to the proteinuria which results from damage to
the glomerular basement membrane of the kidney. This has been
demonstrated in at least one study (GEIGER et al., 1971), in
which increased levels of hPL were noted in the cord blood from
pregnancies complicated by placental insufficiency.

1.4 All Placental Proteins Are Analogues of Proteins Present in
the Normal Adult

In most cases in which a specific placental protein has been
thoroughly characterised it has been found to be analogous to a
product of the normal adult, i.e. sharing some chemical, immunochemical and biological features (Table 1.1); with smaller peptides such as ACTH and the gonadotrophin-releasing hormone there
is no evidence for any difference from the adult product. Excel-
lent examples of "analogues" are provided by the protein hormones
(hCG and hPL) and the alkaline phosphatase enzymes.

With most of the more recently discovered placental proteins
maternal analogues have not, as yet, been defined. This does not
mean that the adult analogue does not exist, merely that it has
not been adequately sought. Equally, it does not exclude the
possibility that some of the placental proteins are completely
unique. The hypothesis of universal analogy is put forward here,
not because it is necessarily true, but because it should be a
stimulus for further work; a placental protein cannot be described as unique until a very extensive search has been made
for comparable material in the adult. Such evidence is not available for any of the proteins in the "analogue unknown" category
in Table 1.1.

Table 1.1. Proteins produced by the human placenta

Protein	Abbreviation	Analogue in non-pregnant adult	No. of subunits	Mol. wt. (Daltons)	Half-life	References
Human chorionic gonadotrophin	hCG	Luteinising hormone	2	45,000–50,000	12–36 h	a,b,c
Human placental lactogen (Human chorionic somatomammotrophin)	hCS	Prolactin, growth hormone	1	21,000–23,000	15–20 min	d
Human chorionic thyrotrophin	hCT	Thyroid stimulating hormone	2	45,000	1	f
Human Chorionic corticotrophin	hCCT	Andrenocorticotrophic hormone	1	5,000	–	g
Human chorionic gonadotrophin releasing hormone	hC-LRH	Gonadotrophin releasing hormone	1	1,000	–	h
Schwangerschafts-spezifisches β_1 glykoprotein	SP_1	Unknown	1	90,000–110,000	30 h	i
Pregnancy specific β_1 glycoprotein	PSβG					j
Trophoblast specific β_1 globulin	TBG					k
Pregnancy associated plasma protein C	PAPP-C					l
Pregnancy associated plasma protein A	PAPP-A	Unknown		750,000	24–48 h	l,m
Pregnancy associated plasma protein B	PAPP-B	Unknown		1,000,000	–24 h	l

Table 1.1. (continued)

Protein	Abbreviation	Analogue in non-pregnant adult	No. of subunits	Mol. wt. (Daltons)	Half-life	Reference
Heat stable alkaline phosphatase	HSAP	Alkaline phosphatase				n,o
Cystine aminopeptidase (Oxytocinase)	CAP	Aminopeptidases				p
Diamine oxidase (Histaminase)	DO	Histaminase	2	190,000		q
Placental protein 5	PP 5	Unknown	1	42,000	15-30 min	r,s

a Bahl et al., 1973.
b Morgan et al., 1973.
c Vaitukaitis, 1977.
d Josimovich, 1977.
e Li, 1970.
f Hershman and Starnes, 1971.
g Opsahl and Long, 1951.
h Gibbons et al., 1975.
i Bohn, 1971.
j Towler et al., 1976.
k Tatarinov and Masyukevich, 1970
l Lin et al., 1976.
m Lin et al., 1976.
n Boyer, 1961.
o Fishman and Ghosh, 1967.
p Tuppy and Nesvadba, 1957.
q Lin and Kirley, 1976.
r Bohn, 1976.
s Obiekwe et al., 1978.

1.5 All Placental Proteins Are Produced in Small Amounts in Normal Adults

This is perhaps the most tenuous of the current set of hypotheses and is based on four pieces of evidence. First, that with at least one specific fetal protein, alphafetoprotein, production is definitely known to extend into adult life, albeit at low levels (PURVES and PURVES, 1972). Second, human chorionic gonadotrophin (hCG) has been isolated from normal testes (BRAUNSTEIN et al., 1975) and pregnancy specific β_1 glycoprotein (SP$_1$) is secreted by fibroblasts in culture (ROSEN et al., 1978) Third, it is the experience of most groups performing sensitive radio-immunoassays for placental proteins that the occasional sample from a normal non-pregnant subject gives a positive result. Usually the result is at or close to the minimum detection limit of the assay, and it is therefore difficult to exclude non-specific effects; yet the possibility exists that these are true positives and that they represent the upper end of a normal range, most of which lies below the detection limit of the assay (Fig 1.4). Fourth, the synthesis of specific placental proteins by tumours indicates that the potential for their production extends into adult life, and "stem-cell" theories of the origin of cancer would imply that low-level production could be continuous.

The value of this hypothesis is that it should stimulate the search for more sensitive techniques of measurement, capable of determining the normal non-pregnant range, and thereby of examining deviations in non-neoplastic pathology.

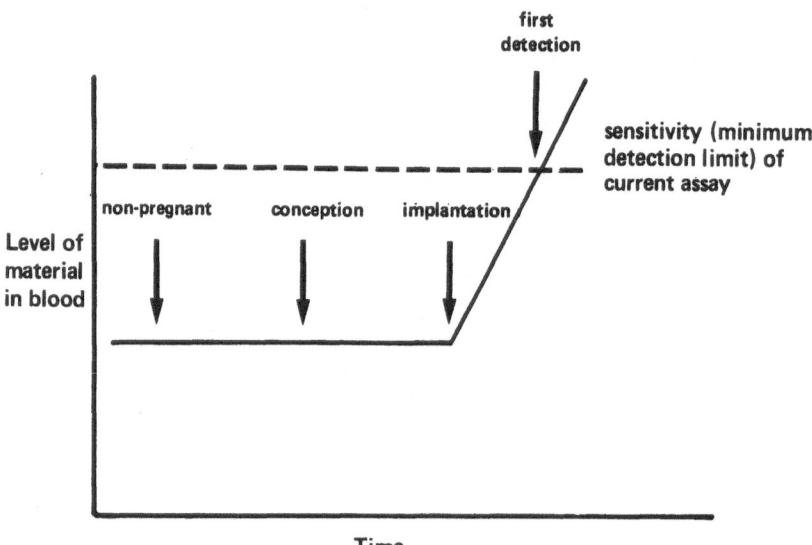

Fig. 1.4. All placental proteins are present in low concentrations in the blood of normal subjects, undetected by current assays. The level rises from the moment of implantation of the blastocyst and the time at which it becomes detectable depends on the sensitivity of the assay used

1.6 The Production Rate of Placental Proteins Is Solely Related to Trophoblast Mass and Blood Flow in the Intervillous Space

It has proved extremely difficult to demonstrate any mechanism which controls the production of any of the specific placental proteins. Where such mechanisms have been suggested (as for example the control of hPL release by maternal carbohydrates and lipids) the evidence is often disputed and subject to alternative explanations (PAVLOU et al., 1973). We have proposed a hypothesis (CHARD, 1976) which is susceptible to experimental study and not at variance with any of the accepted facts about placental protein synthesis. This hypothesis suggests that the potential for placental protein production is a direct function of the total mass of the trophoblast, that the rate of release (and secondarily, of synthesis) is a function of the concentration in the maternal blood in the intervillous space surrounding the syncytiotrophoblast, and that this, in turn, depends on the rate of blood flow in the intervillous space.

This is not the easiest of concepts and can perhaps be clarified by reference to a specific situation. Suppose that an equilibrium is set up with a constant amount of placental tissue, a constant rate of blood flow and, therefore, a fixed concentration of protein in the blood. Now suppose that the blood flow is doubled. Almost immediately the protein concentration will halve because the same amount of hormone is now entering what is effectively twice the volume of blood. According to the hypothesis, the rate of release of the protein will now increase until the concentration returns to its original level. At some time the rate of synthesis will also increase to meet the demand, and a new equilibrium will be established. The situation is analogous to that of dialysis of salts across a semi-permeable membrane; in a closed system solute will flow until the concentrations on the two sides of the membrane are equal; if the solvent on one side of the membrane is replaced solute will again flow until equilibrium is established.

An important implication of this hypothesis is that the rate of placental protein production will be closely related to uteroplacental blood flow. The observed reduction of fetoplacental products in maternal blood in complicated pregnancies could as well be due to this factor as to specific damage to the trophoblast itself. This may go some way to explaining why products of the fetus, such as oestriol, and products of the placenta, such as hPL, often show a concomitant decrease in abnormal states.

It may also explain the rather enigmatic observation that non-placental areas of the trophoblast (i.e. chorion) produce insignificant quantities of placental proteins. This would be a direct result of the fact that such areas, unlike the placental trophoblast, are not directly exposed to a large and fast flowing pool of blood.

1.7 Placental Proteins Have No Biological Function

It is a fundamental tenet of biology that if a structure or material exists, it exists for a purpose, and this philosophy has been extended to the specific placental proteins despite a remarkable paucity of evidence that they serve any function at all. Detection of the biological activity of a protein depends on the existence of a specific assay system, e.g. the luteotrophic activity of hCG, the cleavage of a substrate by an enzyme or the binding of an antigen by an immunoglobulin. However, the precise biological activity of many proteins including albumin, present in high concentrations in the circulation, has not been elucidated. This paucity of information may be due to a lack of experimental models to measure the direct or indirect biological activity of proteins in or on the living cell.

The postulated functions of the placental proteins are of three types: endocrine, metabolic, and immunological. The best-recognised endocrine functions of the placenta are those associated with its production of steroid hormones and their role in the maintenance of pregnancy. Of the protein hormones, hCG has been put forward as the principal luteotrophic factor of early pregnancy (JOHANSSON and BOSU, 1974) and as a possible stimulant of the primitive testis (REGES et al., 1974). The evidence, however, lacks one of the main props of any postulated endocrine activity. It has never been shown that deficiency of the hormone alters its proposed function (i.e. maintenance of the pregnancy or male differentiation). It is extremely difficult, however, to design a set of experiments to test the biological role of hCG in human pregnancy. Immunisation against the β-subunit of hCG is an effective contraceptive but whether this is on the basis of specific inactivation of the biologically active hCG molecule or a non-specific complement effect on the trophoblast itself, has not been proven. In early pregnancy failure the levels of hCG are usually depressed, but it is impossible to know whether this is cause or effect. Furthermore, it has never been demonstrated that administration of hCG can affect the outcome of a normal or abnormal early gestation.

The metabolic effects of the placental protein hormones are mainly attributed to hPL, and include the so-called 'diabetogenic' influence on carbohydrate and lipid metabolism which may benefit the fetus by making more glucose available for placental transfer. Yet, again, there is no solid evidence that hPL is essential for this function (other perfectly valid candidates include the placental steroids), nor is there any evidence that a pregnancy would be any less satisfactory in the absence of these changes. Similarly, the supposed lactogenic function could equally well be attributed to maternal pituitary prolactin, and efficient lactation can be established in a woman who has never been pregnant.

An immunosuppressive role has frequently been suggested for the placental proteins, and consequently that this is the key to the privileged status of the fetus as a graft containing paternal as well as maternal antigens. The published evidence on this role is highly ambiguous; both positive and negative results have been reported when placental proteins are applied to diverse experimental systems, and equally good mechanisms have been proposed which do not involve any specific product of the placenta.

Another function of placental proteins may be their ability to bind other molecules, analogous to the binding of steroids by sex-hormone binding globulin. The advantage of the binding theory is that it can be assessed experimentally in vitro, and could go some way to explain the presence of placental specific proteins during pregnancy.

A search for specific function might have been meaningful when only two or three materials were under consideration. Now, however, the list has extended to a dozen or more and the task becomes Herculean. We propose here a hypothesis which, if accepted, would greatly simplify this area of research: that none of the specific placental proteins has any biological function in the sense of being essential for the mother or the fetus. They are by-products of a more fundamental process concerned with the basic functioning and maintenance of the placenta as an individual and independent organism. In other words, production of placental proteins may indicate nothing other than that the placenta is there. It is also possible that the placental proteins act as a group rather that individually in maintaining the placenta. To date the experimental approach to elucidating biological function has concentrated on assessing purified proteins which may have been denatured during purification, and in future approaches may well be directed to the study of combinations of molecules.

1.8 All Placental Proteins Are Produced by All Tumours

A careful search using sensitive assays is increasingly revealing production of specific placental proteins by a wide range of non-trophoblastic tumours (Table 1.2). Apparently random findings of this type are now so common that they can probably be made the subject of a new hypothesis: that all tumours can produce all placental proteins; and that whether or not an individual protein is found in a particular case depends upon the sensitivity of the assay system used.

The reason for the production of placental (and fetal) antigens by tumours has been the subject of much speculation but no conclusions.Principal theories include the existence in the normal subject of a small group of stem cells, capable of placental protein synthesis, which proliferate in association with the process of neoplasia; and a general reversion of what were otherwise normal cells (e.g. bronchial epithelium) to an embryonic state. Obvious, but so far unhelpful, is the fact that some aspect of a tumour must be comparable to the biology of the normal placenta, and this extends both to production of proteins and to their invasive properties. Based on the observations already made in this chapter, we would like to present a new speculation. Release of placental proteins by trophoblast depends on the existence, in

Table 1.2. Circulating levels of hCG, hPL and SP_1 in subjects with malignant tumours

Malignant tumour	Percentage of subjects with measurable levels		
	hCG	hPL	SP_1
Breast	13-21[a,b]	14[f]	2-22[h,j]
Lung	10[b]	5[d]	3-12[h,j]
Bowel	18[b]	2[d]	20-32[h,i,j]
Ovarian (epithelial)	5-50[b,c,d,e]	21-76[c,d,g]	12-17[c,h,i,j]
Testis	51-62[b,d]	-	11[j]
Melanoma	9[b,d]	-	0[i]

[a] Sheth et al., 1974.

[b] Vaitukaitis et al., 1976.

[c] Crowther et al., 1978.

[d] Rosen et al., 1975.

[e] Stone et al., 1977.

[f] Sheth et al., 1977.

[g] Samaan et al., 1976.

[h] Searle et al., 1978.

[i] Tatarinov and Sokolov, 1977.

[j] Grudzinskas et al., 1978.

immediate contact with the syncytium of trophoblast cells, of a fast moving stream of blood; hence the high level of production in the placenta itself, the very low level of production in the non-placental trophoblast (i.e. the chorion), and the absence of significant transfer to the fetus. In an adult tumour we would propose that there are cells, or groups of cells, which because of their invasive properties and syncytial nature are in direct contact with the blood stream, and not separated from it by endothelium and basement membrane. It is this property of direct contact with blood, and the potential of the cell to synthesise the proteins, which explains the production of placental proteins by adult tumours.

There are a number of observations which conflict with the theory of direct contact stimulating placental protein production. First, the relative concentration of placental proteins in the circulation of cancer patients often differs from that present in the maternal blood during pregnancy (CROWTHER et al., 1978). Second, the biochemical properties of hCG isolated from tumours differs from that found in normal pregnancy (WEINTRAUB et al., 1975) and unbalanced or isolated ectopic production of the subunits of hCG has also been demonstrated in certain tumours. Third, many highly vascular tumours do not produce measurable levels of placental proteins. Fourth, endothelial cells lining normal blood vessels do not produce placental proteins. However, direct cell-blood contact may be important in stimulating the production rate of placental proteins in cancer cells which possess the synthetic mechanism.

1.9 Production of Placental Proteins by Tumours Is as Irrelevant as Production by the Placenta

It has never been suggested that the production of placental proteins and peptides by a tumour serves any essential biological purpose though they may, of course, have biological effects (i.e. Cushing's syndrome). This hypothesis could, therefore, be described as trivial were it not that it reflects on the same hypothesis already made for protein production by the placenta. Here, we would re-emphasise our proposal that production of specific proteins by the placenta is as meaningless in functional terms as is ectopic hormone secretion by tumours. Both are by-products of some fundamental but ill-understood process, and both may depend on the relationship of particular types of cells to their vascular supply.

1.10 Levels of Placental Proteins in Maternal Blood Parallel the Growth of the Placenta

Where adequate information is available, the concentration of specific placental proteins in maternal blood have, with one exception, been shown to follow a sigmoid curve of the type shown in Fig. 1.5; a relatively slow increase in early pregnancy, a rapid rise in the midtrimester, a plateau from the 36th week onwards, and possibly a fall after term. This pattern parallels exactly the growth of the placenta, whether assessed by weight (HYTTEN and LEITCH, 1971), or by DNA content. Furthermore, it is a logical conclusion of our earlier hypothesis that the production of placental proteins depends solely on the total mass of the trophoblast and the blood flow in the intervillous space; the latter would, in a group of normal pregnancies, be expected to follow fairly closely the overall growth of the placenta. The curve of placental protein concentration throughout pregnancy has a sigmoid shape when plotted on an arithmetic scale but this conceals the exponential rise in concentration when the levels are plotted on a logarithmic scale (Fig. 1.5). The logarithmic plot shows that the rate of increase in concentration is maximal in early pregnancy and decreases with time. This makes the assessment of falling levels in an individual subject extremely difficult if not impossible.

The obvious exception to this pattern is hCG, which reaches a massive peak in the later part of the first trimester and thereafter declines rapidly to follow a "normal" curve in the late second and third trimesters. At first sight it could be said that the observation of this pattern negates any attempt to generalise on the overall trend of placental proteins in maternal blood. However, we would propose that the first trimester peak of hCG is so aberrant, and the arguments for a sigmoid pattern of all other proteins so cogent, that "first trimester hCG" should be regarded as an altogether separate and unique material which probably has a different cellular origin and different control mechanisms to all other placental proteins (Fig. 1.6). This proposal could be subjected to experimental test which has not, as yet, been carried out; differences in physicochemical properties and immunohistochemical localisation between "first" and "third" trimester hCG would be obvious lines to explore.

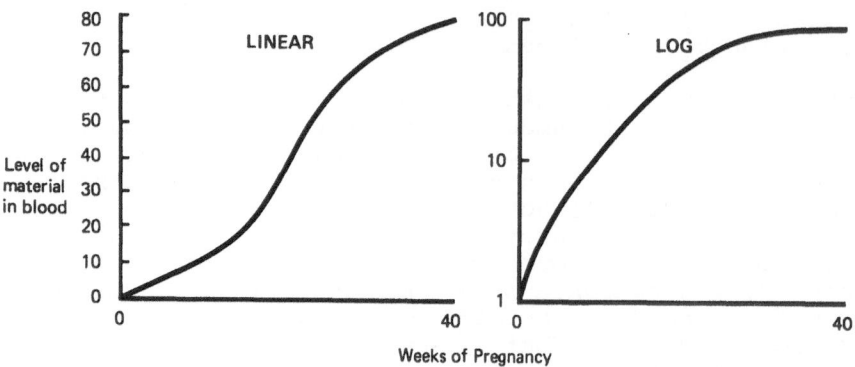

Fig. 1.5. Circulating concentrations of a typical placental protein product throughout pregnancy, plotted on an arithmetic and a logarithmic scale. The latter shows that the greatest rate of increase is in early pregnancy

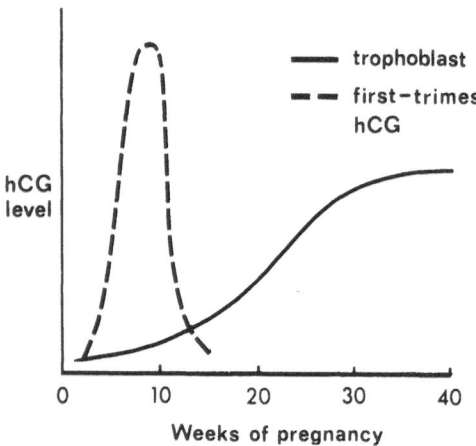

Fig. 1.6. The pattern of hCG levels in a normal pregnancy. hCG from the syncytiotrophoblast shows the sigmoid pattern characteristic of all specific placental proteins. The first trimester peak is a variant and must be due to a completely different site and mechanism of synthesis

1.11 The Clinical Significance of Maternal Blood Levels Is Identical for All Placental Proteins

Nothing that has been stated in previous sections (origin, control mechanisms, functions) would lead one to expect any notable difference in the clinical significance of the circulating placental proteins. If, in a compromised pregnancy, one is decreased all will be decreased. Any variation could be attributed only to the precision and practicality of measurement of an individual protein.

It has often been claimed that variations in the half-life of different placental proteins will alter their usefulness as a clinical marker (KLOPPER, 1976). However, if the rate of release is controlled by the concentration of blood flowing through the intervillous space, then provided the half-life is longer than a few minutes, the concentrations of all the proteins will remain constant. Diurnal and day-to-day variation studies on hPL and SP1, two proteins which vary in size, (hPL, 20,000; SP1, 100,000) and half-lives (hPL, 15 min; SP1 30 h), have shown that the variation is virtually identical, thus adding weight to this hypothesis (MASSON et al., 1977). It is further claimed that a short half-life confers greater intrinsic value to the test because the circulating levels will more rapidly reflect an acute event such as placental abruption. But this situation is rare and biochemical tests are of little practical value; most complications of late pregnancy which threaten the well-being of the fetus are part of a relatively long-term process (GUSTAVSSON et al., 1977).

In our opinion there is nothing in the available experimental evidence which vitiates the very simplistic hypothesis that assays of all placental proteins have the same value. The literature is replete with claims and counter-claims, but positive results, when put forward, always have a virtually identical clinical interpretation. We would propose, therefore, that the clinical significance is in reality the same for all placental proteins, and that the choice of one or the other must ultimately rest on practical considerations of the method of assay.

It would be appropriate at this point to re-emphasise some general aspects of the execution and interpretation of placental function tests which we have already presented elsewhere (CHARD, 1976; GORDON and GRUDZINSKAS, 1978).

In late pregnancy any test of fetal wellbeing must answer two questions. First, is the test able to pinpoint those pregnancies in which the fetus has an increased risk of dying or developing perinatal asphyxia or growth retardation, and is the test a better predictor than other clinical, biochemical or biophysical measurements currently available? Second, can the result be used to decide on the optimal time for delivery of the baby? The answer to the first question is complicated by the lack of an objective method of assessing the baby after delivery because of the multifactorial aetiology of perinatal morbidity and mortality. No test of placental function will predict a traumatic delivery or cord prolapse and the majority of cases of severe mental retardation in infancy are due to events which occurred during the early weeks of pregnancy such as chromosomal abnormalities or maldevelopment of the neural tube (GUSTAVSSON et al., 1977) Thus only a small proportion of perinatal morbidity is due to placental dysfunction in late pregnancy (Fig. 1.7). It is nevertheless possible to assess the value of a placental function test and to compare it with other parameters, provided an entire obstetric population is surveyed and data is available for the majority of subjects (LETCHWORTH et al., 1978; GORDON et al., 1978). The advantage of a global survey is that it forms the basis for assessing true and false positive and negative results, and the ratio of true positives to false negatives provides the basis from which the relative risk statistic can be calculated. The results of one prospective survey in which this method was used showed that in comparative terms depressed circulating levels of hPL were associated with a higher risk of neonatal morbidity than any other antenatal parameter tested,

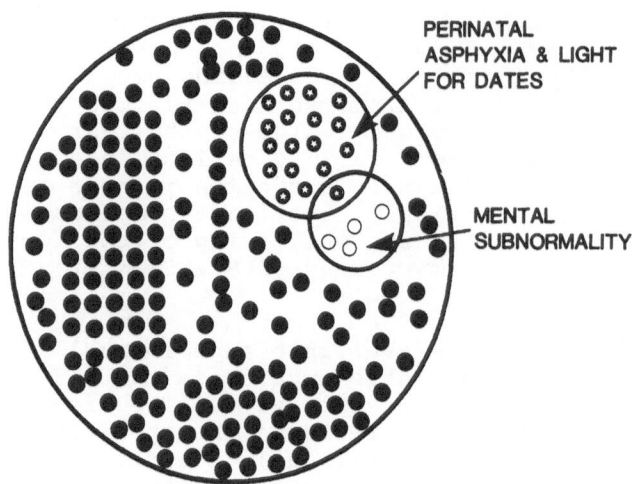

Fig. 1.7. A hypothetical obstetric population. A small proportion of the children will be mentally subnormal. Some of these may be associated with perinatal asphyxia or intrauterine malnutrition. The majority are associated with other abnormalities (eg. trisomy 21) which cannot be readily identified in late pregnancy

except in severe pre-eclampsia (Table 1.3) (GORDON et al., 1978). However the hPL results were not compared with ultrasonic measurement of the fetus or with antenatal fetal heart monitoring in late pregnancy. It is possible that the prediction of fetal risk using ultrasonography will be superior because the fetus is assessed directly, whereas hPL is an indirect index. (GORDON and GRUDZINSKAS, 1978). The second question, about the use of placental function tests in the timing of the delivery of the baby, can only be answered by means of a controlled trial in which half the results are reported and the remainder analysed after delivery. There have been two such studies. In one by SPELLACY and his colleagues (1975) using hPL the conclusion was that the test was of value. In our experience, however, subjects with low levels of hPL do not show a fall and this is consistent with the exponential nature of the rate of increase in circulating concentrations in late pregnancy. The hPL values can only be used indirectly in the timing of induction. In the second controlled study in which plasma oestriol levels were used as a marker of fetal wellbeing in late pregnancy, a significant number of false positive results occurred and the conclusion reached by the authors was that the test was not valuable in timing delivery (DUENHOELTER et al., 1976). In view of the high false positive rate of placental protein assays in the prediction of a compromised fetus it is important that they are only used as a screening test for fetal risk. The clinical use of many of the other placental proteins has not been subjected to critical analysis to date and the results will be of great interest. However, until this data is available the ease and accuracy of measurement is probably the most important criterion in judging which placental protein to use in clinical practice (CHARD, 1976).

Table 1.3. Relative risk as an index of fetal wellbeing

Parameters studied	Relative risk
hPL levels after 30 weeks >90th centile	0.2[a]
Mild essential hypertension (diastolic blood pressure 90-110 mm Hg)	0.5
Average weekly weight gain (20-40 weeks) >90th centile	0.5
Maternal weight at 32 weeks >90th centile	0.6
Non-smokers	0.6
Threatened abortion	0.6
Average weekly weight gain (20-40 weeks) <10th centile	0.7
Previous perinatal death (multiparous subjects)	0.7
Maternal age >20 years	1.0
Mild pre-eclampsia (diastolic blood pressure 90 mm Hg, no albuminuria)	1.0
Biparietal diameter (weeks 18-22) <10th centile	1.1
Non-Caucasian ethnic origin	1.3
Elevated maternal AFP levels (weeks 16-20)	1.3
Antepartum haemorrhage	1.3
Maternal cardiac, renal, gastrointestinal diseases	1.3
Meconium stained amniotic fluid in labour	1.7
Maternal weight at 32 weeks <10th centile	1.8
Smoking >15 per day at booking	2.0[b]
Maternal age >34 Years	2.4[b]
hPL levels after 30 weeks <10th centile	2.9[c]
Fetal heart abnormality (in first stage of labour)	3.2[c]
Severe pre-eclampsia and essential hypertension	8.6[c]

Chi-squared test: significant at 5% level[a], 1% level[b], 0,1% level[c].

1.12 Conclusions

We have put forward a series of hypotheses on the placental proteins. Some of these will prove contentious, but are intended as a stimulus to discussion and possibly to further experimental work. With others, we would place the onus on our colleagues to demonstrate that the proposal is fundamentally wrong: for example, with a claim that a particular protein is biologically unique, that its function is meaningful, or that its clinical significance is pre-eminent over all others. Simplification is essential if this fascinating subject is not to degenerate into a chaos of isolated and conflicting views.

References

Bahl, O.P., Carlsen, R.B., Bellisario, R.: Human chorionic gonadotrophin – amino acid sequence of the α and β subunits. Biochem. Biophys. Res. commun. 48, 416-422 (1973)

Bennett, M.J., Grudzinskas, J.G., Gordon, Y.B., Turnbull, A.C.: Circulating levels of alpha-fetoprotein and pregnancy specific β_1 glycoprotein in pregnancies without an embryo. Br. J. Obstet. Gynaecol. 85, 348-349 (1978)

Bohn, H.: Nachweis und Charakterisierung von Schwangerschafts-Protein in der menschlichen Placenta, sowie ihre quantitative immunologische Bestimmung im Serum schwangerer Frauen. Arch. gynaekol. 210, 440-457 (1971)

Bohn, H.: Isolation and Characterization of placental specific proteins SP_1 and PP5. In: Protides of the biological fluids. PEETERS, H. (ed.), pp. 117-124. Oxford: Pergamon Press 1976

Boyer, S.H.: Alkaline phosphatase in human serum and placentae. Science 134, 1002-1016 (1961)

Braunstein, G.D., Rasor, J., Wade, M.E.: Presence in normal human testes of a chorionic gonadotrophin-like substance distinct from human luteinizing hormone. N. Engl. J. Med. 293, 1339-1343 (1975)

Burgos, M.H., Rodriguez, E.M.: Specialized zones in the trophoblast of the human term placenta. Am. J. Obstet. Gynecol. 96, 342-356 (1966)

Chard, T.: Normality and abnormality. In: Plasma hormone assays in evaluation of fetal wellbeing. KLOPPER, A. (ed.), pp. 1-11. Edinburgh: Churchill Livingstone 1976

Crowther, M.E., Grudzinskas, J.G., Poulton, T.A., Gordon, Y.B.: The measurement of circulating levels of pregnancy specific β_1 glycoprotein, human chorionic gonadotrophin and human placental lactogen in non-trophoblastic carcinoma of the ovary. Am. J. Obstet. Gynecol. (1978) (in press)

Currie, A.R., Beck, J.S., Ellis, S., Read, C.H.: Immunofluorescent localisation of a growth-hormone-like factor in normal and abnormal syncitiotrophoblast. J. Pathol. Bacteriol. 92, 395-399 (1966)

Duenhoelter, J.H., Whalley, P.J., MacDonald, P.C.: Analysis of the utility of plasma immunoreactive estrogen measurements in determining the delivery time of gravidas with a fetus considered to be at high risk. Am. J. Obstet. Gynecol. 125, 889-898 (1976)

Edwards, R.G.: Early human development: from oocyte to implantation. In: Scientific foundations of obstetrics and gynaecology, 2nd ed. PHILIP, E.A., BARNES, J., NEWTON, M. (eds.), pp. 175-252. London: Heinemann 1977

Fishman, W.H., Ghosh, N.K.: Isoenzymes of alkaline phosphatase. Adv. Clin. Chem. 10, 255-270 (1967)

Fox, H.: The incidence and significance of vasculo-syncitial membranes in the human placenta. J. Obstet. Gynaecol. Brit. Common. 74, 28-34 (1967)

Gau, J., Chard, T.: The distribution of placental lactogen in the human term placenta. Brit. J. Obstet. Gynaecol. 82, 790-793 (1976)

Geiger, W., Kaiser, R., Franchimont, P.: Comparative radioimmunological determination of human chorionic gonadotrophin in foetal and maternal blood after delivery. Acta Endocrinol. (Kbh) 68, 169-182 (1971)

Getzowa, S., Sadowsky, A.: On the structure of the human placenta with full time and immature foetus, living or dead. J. Obstet. Gynaecol. Br. Emp. 57, 388-396 (1950)

Gibbons, J.M., Mitnick, M., Chieffo, V.: In vitro biosynthesis of TSH and LH releasing factors by the human placenta. Am. J. Obstet. Gynecol. 121, 127-131 (1975)

Gitlin, J.D., Gitlin, D.: Cell receptors and the selective transfer of proteins from mother to young across tissue barriers. Pediat. Res. 7, 290-298 (1973)

Gordon, Y.B., Grudzinskas, J.G.: The role of biochemical tests and ultrasound in antenatal diagnosis. In: Diagnostic ultrasound applied to obstetrics and gynaecology. SABBAGHA, R. (ed.). New York: Harper and Row 1978 (in press)

Gordon, Y.B., Lewis, J.D. Pendlebury, D.J., Leighton, M., Gold, J.: Is measurement of placental function and maternal weight worthwhile? Lancet 1000-1003 (1978)

Grudzinskas, J.G., Coombes, R.C., Ratcliffe, J.G., Gordon, Y.B., Munro Neville, A., Chard, T.: Circulating levels of pregnancy specific β_1 glycoprotein in patients with testicular, bronchogenic carcinomas. (In preparation) (1978)

Gustavsson, K.H., Hagberg, B., Hagberg, G., Sars, K.: Severe mental retardation in a Swedish county. Neuropädiatrie 8, 293-304 (1977)

Hamilton, W.J., Hamilton, D.V.: Development of the human placenta. In: Scientific foundations of obstetrics and gynaecology, 2nd ed. PHILIP, E.A., BARNES, J., NEWTON, M. (eds.), pp 292-358. London: Heinemann 1977

Hershman, J.M., Starnes, W.R.: Placental content and characterization of human chorionic thyrotrophin. J. Clin. Endocrinol. Metab. 32, 52-57 (1971)

Horne, C.H.W., Towler, C.M., Pugh-Humphreys, R.G.P., Thomson, A.W., Bohn, H.: Pregnancy specific β_1 glycoprotein - a product of the syncitiotrophoblast. Experientia 32, 1197-1199 (1976)

Hytten, F.E., Leitch, I. (eds.): Physiology of human pregnancy, pp. 326-329. Oxford: Blackwell Scientific 1971

Johanssen, E.D.B., Bosu, W.T.K.: The function of the corpus luteum in primates. In: Physiology and genetics of reproduction, Part B. COUTINHO, E.M., FUCHS, F. (eds.), pp. 343-352. New York: Plenum Press 1974

Josimovich, J.B.: Human placental lactogen. In: Endocrinology of pregnancy, 2nd ed. FUCHS, F., KLOPPER, A.(eds.), pp. 191-205. New York: Harper and Row 1971

Klopper, A.: Criteria for the selection of steroid assays in the assessment of fetoplacental function. In: Plasma hormone assays in evaluation of fetal wellbeing. KLOPPER, A. (ed.), pp. 20-35. Edinburgh: Churchill Livingstone 1976

Letchworth, A.T., Slattery, M., Dennis, K.J.: Clinical application of human placental lactogen values in late pregnancy. Lancet 955-957, 1978

Li, C.H.: On the characterization of human chorionic somatomammotropin. Ann. Sclavo 12, 651-659 (1970)

Lin, Chi-wei, Kirley, S.D.: Human placental and tumour histaminase. In: Protides of the biological fluids. PEETERS, H. (ed.), pp. 103-108. Oxford: Pergamon Press 1976

Lin, T.M., Halbert, S.P.: Placental localization of human pregnancy associated plasma proteins. Science (N.Y.) 193, 1249-1252 (1976)

Lin, T.M., Halbert, S.P., Kiefer, D., Spellacy, W.N., Gall, S.: Characterization of four human pregnancy-associated plasma proteins. Am. J. Obstet. Gynecol. 118, 223-236 (1974)

Lin, T.M., Halbert, S.P., Spellacy, W.N., Gall, S.: Human pregnancy-associated plasma proteins during the post-partum period. Am. J. Obstet. Gynecol. 124, 382-387 (1976)

Masson, G.M., Klopper, A.I., Wilson, G.R.: Plasma estrogens and pregnancy-associated plasma proteins: A study of their variability. Obstet. Gynecol. 50, 435-438 (1977)

Morgan, F.J., Birken, S., Canfield, R.E.: Human chorionic gonadotrophin: A proposal for its amino acid sequence. Mol. Cell. Biochem. 2, 97-99 (1973)

Obiekwe, B., Grudzinskas, J.G., Gordon, Y.B., Bohn, H., Chard, T.: Circulating levels of placental protein 5 (PP5) in the third trimester of pregnancy. Proceedings of 6th meeting, international research group for carcino embryonic proteins. 152, 1978

Opsahl, J.C., Long, C.N.H.: Identification of ACTH in human placental tissue. Yale J. Biol. Med. 24, 199-209 (1951)

Pavlou, C., Chard, T., Landon, J., Letchworth, A.T.: Circulating levels of human placental lactogen in late pregnancy: The effect of glucose loading, smoking and exercise. Eur. J. Obstet. Gynaecol. Reprod. Biol. 3, 45-49 (1973)

Purves, L.R., Purves, M.: Serum alpha-fetoprotein. S. Afr. Med. J. 46, 1290-1296 (1972)

Reyes, F.I., Boroditsky, R.S., Winter, J.D.S., Faiman, C.: Studies on human sexual development. J. Clin. Endocrinol. Metab. 38, 612-617 (1974)

Rosen, S.W., Weintraub., B.D., Vaitukaitis, J.L., Sussman, H.H., Hershman, J.M., Muggia, F.M.: Placental proteins and their subunits as tumour markers. Ann. Intern. Med. 82, 71-83 (1975)

Rosen, S.W., Kaminska, J., Calvert, I.S., Aronson, S.: Human fibroblasts produce $PS\beta_1G$ in vitro. Proceedings of 6th meeting, international research group for carcino-embryonic proteins 147 (1978)

Samaan, N.A., Smith, J.P., Rutledge, F.N.: The significance of measurement of human placental lactogen, human chorionic gonadotrophin and carcino-embryonic antigen in patients with ovarian carcinoma. Am. J. Obstet. Gynecol. 126, 186-189 (1976)

Sciarra, J.J., Kaplan, S.L., Grumbach, M.M.: Localization of anti-human growth hormone serum within the plasma: Evidence for a human chorionic growth hormone prolactin. Nature 199, 1005-1006 (1963)

Searle, F., Leake, B.A., Bagshawe, K.D., Dent, K.: Serum-SP_1-pregnancy specific β_1 glycoprotein in choriocarcinoma and other neoplastic diseases. Lancet I, 579-580 (1978)

Sheth, N.A., Suraiya, J.N., Ranadive, K.J. Sheth, A.R.: Ectopic production of human chorionic gonadotrophin by human breast tumours. Br. J. Cancer 30, 566-570 (1974)

Sheth, N.A., Suraiya, J.N., Sheth, A.R., Ranadive, K.J., Jussawalla, D.J.: Ectopic production of human placental lactogen by breast tumours. Cancer 39, 1693-1699 (1977)

Silman, R.E.; Why babies are born. (In preparation) (1978)

Spellacy, W.N., Buhi, W.C., Birk, S.A.: The effectiveness of human placental lactogen measurements as an adjunct in decreasing perinatal deaths. Am. J. Obstet. Gynecol. 121, 835-844 (1975)

Stone, M., Bagshawe, K.D., Kardana, B.: β-human chorionic gonadotrophin and carcinoembryonic antigen in the management of ovarian carcinoma. Br. J. Obstet. Gynaecol. 84, 375-379 (1977)

Tatarinov, Yu S., Masyukevich, V.N.: Immunological identification of a new beta$_1$-globulin in the blood serum of pregnant women. Byull. Eksp. Biol. Med. 69, 66-68 (1970)

Tatarinov, Yu S., Sokolov, A.V.: Development of a radioimmunoassay for pregnancy specific Beta$_1$-globulin and its measurement in serum of patients with trophoblastic and non-trophoblastic tumours. Int. J. Cancer 19, 161-166 (1977)

Towler, C.M., Horne, C.H.W., Jandial, V., Campbell, D.M., MacGillivray, I.: Plasma levels of pregnancy-specific β$_1$ glycoprotein in normal pregnancy. Br. J. Obstet. Gynaecol. 83, 775-779 (1976)

Tuppy, H., Nesvadba, H.: Über die aminopeptidaseaktivität des Schwangerenserums und ihre Beziehung zu dessen Vermögen, Oxytocin zu inaktivieren. Monatshilfe Chemie 88, 977-983

Vaitukaitis, J.L.: Human chorionic gonadotrophin. In: Endocrinology of pregnancy, 2nd ed. FUCHS, F., KLOPPER, A. (eds.), pp. 63-76. New York: Harper and Row 1977

Vaitukaitis, J.L., Ross, G.T., Braunstein, G.D., Rayford, P.L.: Gonadotrophins and their subunits. Recent Prog. Horm. Res. 32, 289-330 (1976)

Weintraub, B.D., Krauth, G., Rosen, S.W., Rabson, A.S.: Differences between purified ectopic and normal alpha subunits of human glycoprotein hormones. J. Clin. Invest. 56, 1043-1052 (1975)

2 The Measurement of Trophoblastic Proteins as a Test of Placental Function

A. Klopper, R. Smith, and I. Davidson

Ever since TATARINOV and MASYUKEVICH (1970), followed by BOHN (1971) and GALL and HALBERT (1972) adduced evidence that the trophoblast produces and secretes into maternal blood a number of proteins other than chorionic gonadotrophin (hCG) and human placental lactogen (hPL) two questions have been uppermost in the minds of investigators in this field. The first was whether the concentration of these proteins in the maternal circulation gave any indication of the functional state of the placenta and the second was what their physiological function might be.

The use of placental protein assays as a measure of placental function in obstetric disease is an obvious possibility and now that simple means to measure these proteins are available, it is safe to predict that there will be a spate of publications on this topic in the next few years. For the most part these will deal with empirical observations and it will be some time before the limits of these tests are discerned. In truth, the value of these measurements cannot fully be defined until the function of these proteins is known. It is the purpose of this chapter to suggest some criteria by which the value of plasma assays might be assessed; to compare various proteins by these criteria; and to put forward some speculations as to their function. We propose to consider first the pregnancy-specific β_1 glycoprotein designated SP_1 by BOHN (1972) and PAPP-C by LIN et al. (1974).

There has been a tendency to settle on the designation of $PS\beta_1G$ for this protein but this is likely to cause confusion with PAPP-B, which is also a pregnancy-specific glycoprotein with β_1 electrophoretic mobility. As Hans Bohn has pioneered much of the work on this protein it is perhaps appropriate to use his term "Schwangerschaftsprotein" (SP_1) until a functional designation can be evolved. A second candidate we propose to consider is the protein designated PAPP-A by LIN et al. (1974). hPL assays have an established place in the assessment of placental function, and we will use these as a baseline for comparison. Although it is not a protein, oestriol measurements loom so large in the determination of fetal well-being that some comparisons with this steroid are inescapable.

2.1 Methodology

Although it is seldom explicitly stated, simple methods have an attraction well beyond the value of the information they yield. If the new placental proteins can easily be measured, they will be enthusiastically welcomed regardless of how much we understand of their function. And of this there is no doubt: in late pregnancy they can easily be measured without elaborate instrumentation or sophisticated technology. At first SP_1 was measured by immunodiffusion techniques which involved little more than putting a measured volume of serum onto a prepared agar plate, commercially available, and measuring the precipitin ring with a ruler 4 or 5 days later. The method we have used (BRUCE and KLOPPER, 1978) is hardly more elaborate. It is a modification of the standard Laurell "rocket" immunoelectrophoresis technique and can be carried out in any laboratory having a simple, cheap electrophoresis apparatus, electric current and a running water supply. The methodology is simple enough to make the measurement of these placental proteins in late pregnancy available to all.

2.2 Normal Values

A fairly strict and preferably generally accepted definition of the normal range at various stages of gestation is a sine qua non for judging the value of these assays. Except for plasma oestriol concentration there is little change in the mean value of any of these parameters in the last 4 weeks of pregnancy. For the most part the decisions which are liable to be influenced by a particular assay are made around 38 weeks of pregnancy. This analysis is, therefore, centred on this stage of pregnancy.

A great many normal series have been published for hPL. They differ rather more than can be accounted for by the state of the methodology, but the figures in Table 2.1 are close to the median published value and likely to be as close to the truth as any. The position of normal values for SP_1 is less certain. Although a commercial standard exists and the values in Table 2.1 are recorded in terms of that standard, their relationship to absolute weight of protein is an approximation which may have to be revised in the future. There is as yet no accepted standard

Table 2.1. Placental protein concentration at 38 weeks gestagen (mg/litre)

Mean	Standard deviation	Coefficient of variation (%)	Authors
hPL			
5.8	1.2	21	LINDBERG and NILSSON, 1973
6.2	2.2	35	KLOPPER et al., 1977
SP_1			
139.3	42.5	31	TATRA et al., 1974
199.0	53.9	27	TOWLER et al., 1976
159.0	48.0	30	KLOPPER et al., 1977

for PAPP-A and the values recorded in this assay are in terms of an arbitrary pregnancy plasma pool. They are, however, adequate to the purpose of comparison of range and change within the body of data presented there. A noteworthy implication of the figures in Table 2.1 is that the concentration of SP_1 is some 20 times that of hPL. Of course this is a rough approximation, but there is no doubt that in the case of SP_1 we are dealing with amounts of material which far exceed those of any of the substances which constitute the established parameters for the assay of placental function.

Another characteristic of the normal values has some bearing on the use of plasma measurements for assessing placental function. This refers to the shape of the mean curve during pregnancy. hCG peaks at 12 weeks and is so low in the third trimester that its assays have never been used with any success in late pregnancy. α Fetoprotein peaks later but the levels are decreasing during the critical time in late pregnancy. One never, knows, therefore, whether any decline in the concentration of α fetoprotein is due to a failure of placental function or merely reflects a natural decline. The shape of the mean curve for hPL and for SP_1 resembles that of the growth curve of the placenta itself; that is to say it flattens out and remains level over the last 4 weeks of pregnancy. One, therefore, lacks the reassurance of rising values from week to week when monitoring pregnancy which is at risk. Oestriol concentration certainly, and PAPP-A probably, rise right up to term, and by doing so add an extra dimension to these measurements in the assessment of fetal well-being.

2.3 Subject-to-Subject Variability

A problem which has plagued many diagnostic measurements is the large spread of normal values from subject to subject. This is especially true for the parameters of placental function under discussion. We do not know what controls the production of any of the products and have only a few uncertain guesses as to why one normal woman should have a great deal more or less oestriol, hPL, SP$_1$ or PAPP-A than another at the same stage of gestation. The concentration of oestriol is somewhat uncertainly linked to the size of the fetus, and it seems a safe guess that trophoblastic products such as hPL, SP$_1$ and PAPP-A should reflect functional placental mass, but even the most sophisticated analyses have failed to produce any more certain information than that (LIN et al., 1976).

In all four measurements there is an overlap between the lowest normal values and the highest pathological values. The narrower the normal range, the better is the chance of clearly distinguishing normal from abnormal. The usual way of describing the normal range is by calculating the standard deviation about the mean. In order to compare entities measured in widely different units it is necessary to express the standard deviation as a proportion of the mean, i.e. as a coefficient of variation. The coefficients of variation for 60 normal women at 38 weeks gestation are shown in Table 2.2.

The measurements were done in groups of 12 subjects for each parameter and, therefore, the variability figures include the methodological inter-assay variability as well as the physiological variability from subject to subject. The methodological variability for oestriol, which is measured by radioimmunoassay, is higher (12%) than the variability for SP$_1$ measurements (5%), which require fewer and simpler manipulations. This does not, however, account entirely for the difference between oestriol and the other parameters shown in Table 2.2 for hPL was also measured by radioimmunoassay. It appears that the subject-to-subject spread of protein concentration measurements is less than that of the steroids and by this criterion any of the protein measurements are better parameters of placental function than the steroid assays (KLOPPER et al., 1977). By this criterion there is nothing to choose between hPL, SP$_1$ and PAPP-A.

At best all these parameters show an alarmingly wide spread and the possibility of distinguishing clearly between pathological and normal on the basis of a single estimation in an individual is limited. There is a feature, common to the normal distribution of all the values, which somewhat diminishes this disadvantage. Describing the normal range in terms of mean and standard deviation as was done in Table 2.2 assumes that the values are equally distributed about the mean. This is in fact not so. All the parameters have a skewed distribution; the mean is pulled up by a few high values and there are more values below than above the mean. This can be corrected either by logarithmic transformation of the values or by plotting the range in terms of percentiles about the median. Such an analysis is shown for SP$_1$ in Fig. 2.1. Similar findings apply to all other parameters. When the normal range is defined in this way it has the effect of raising the lower limit of normal, and, therefore, sharpens the discrimination between low normal and the highest pathological values. It also raises the upper limit of normal but this is of no clinical interest as the desired area of discrimination is in the low values.

Table 2.2. Subject-to-subject variability of placental proteins and of oestriol

Substance	Coefficient of variation (%)
Total oestriol	44
Unconjugated oestriol	42
hPL	36
SP_1	30
PAPP-A	30

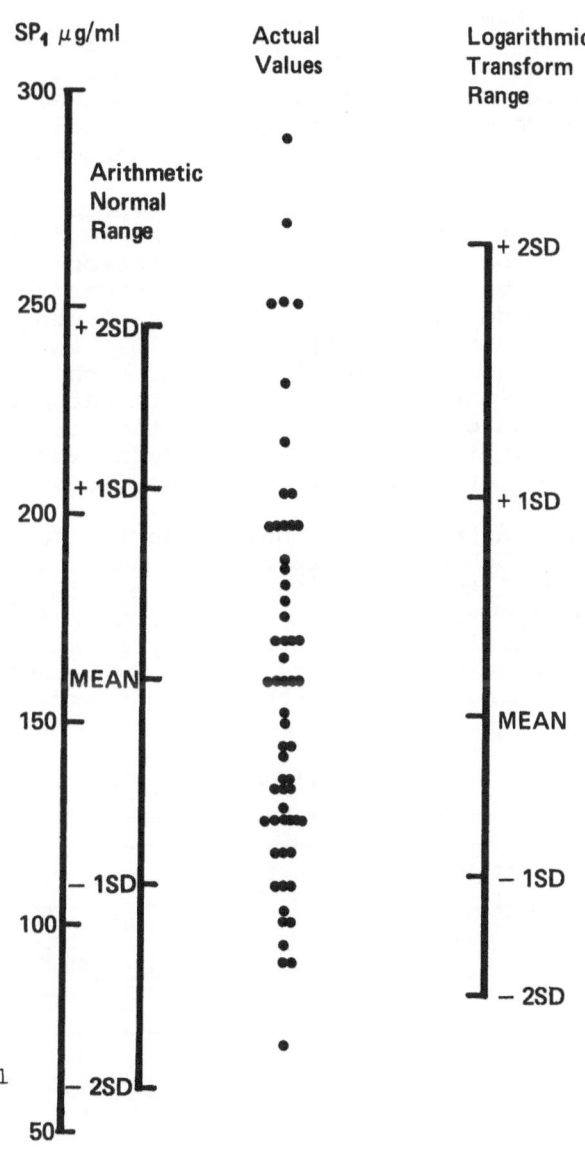

Fig. 2.1. The spread of normal SP_1 values at 38 weeks gestation (n=60)

2.4 Day-to-Day Variation

It is a platitude of placental function measurements, repeated in many publications, that serial assays are much more reliable than single determinations. In part this is because the most widely used measurement, oestriol, is notoriously variable from one time to another. There is a limitation of understanding in this field which cripples experiment: we do not know what factors determine the rate of oestriol production, or indeed that of any of the other parameters. Any, or all of them could be produced at a steady rate, and all of the ups and downs could be produced by alterations in uterine flow. In that case there would be little in terms of time-to-time variability to choose between one parameter and another. On the other hand, the plasma concentration of any substance is the result, not only of inflow of the substance into the maternal plasma compartment but also of its rate of removal from the plasma. We are dealing with a dynamic balance between inflow and outflow; a balance which may change very rapidly and whose alteration need not necessarily be reflecting changes in placental production.

Given the situation where the variables are so poorly understood, the most useful thing which can be done about this important criterion of time-to-time variation is an empirical comparison of the various parameters. In order to do so blood was taken from nine subjects at 38 weeks of pregnancy for 8 consecutive days. In this way it was possible to determine the mean value of each parameter over 8 days for each subject and to calculate the coefficient of variation about that mean for each subject. Taking the mean coefficient of variation of all nine subjects gave an overall figure for the day-to-day variability of each parameter (MASSON et al., 1977). The results are shown in Table 2.3.

Unfortunately PAPP-A determinations were not made in this experiment. Subsequent experiments suggested that the day-to-day variability of PAPP-A was 7%; all the proteins behave roughly the same and are much less variable from time to time than are the steroids. By this second criterion of day-to-day variability all three of the proteins are again better than oestriol and there is nothing to choose between one protein and another.

Table 2.3. Day-to-day variability of placental proteins and oestriol

Substance	Coefficient of variation (%)
Total oestriol	17
Unconjugated oestriol	15
hPL	6
SP_1	5

2.5 Puerperal Decline

A further characteristic of the plasma concentrations of any of
the substances under survey which determines the usefulness of
any particular assay is the speed with which it changes when production of the substance changes. In this respect there is a convenient natural model to hand. This is constituted by delivery
of the fetus and placenta which, at a stroke, cuts off all production of placental protein or steroid. A convenient measure¯
of the post-delivery rate of decline is the half-life, i.e. the
time taken for the plasma concentration to fall by half after
delivery. Studies of the puerperal decline have a bearing on
two aspects of the assessment of placental function; firstly
they show how closely geared to placental function the parameter
is, and secondly they give an insight into the maternal compartments which the substance may enter.

Half-life studies were done on 10 normal women delivering at
term. The mean half-lives are recorded in Table 2.4. The data
were taken from KLOPPER et al. (1978).

It is evident that oestriol and hPL have very short half-lives,
PAPP-A is somewhat longer and SP_1 is very much longer.

The dramatic difference in the initial rate of removal becomes
evident when one considers only the concentrations in the first
hour after delivery. Figure 2.2 shows a comparison between the
mean percentage change in oestriol and in SP_1. After an hour
the concentration of oestriol had fallen to 20% of its initial
level while SP_1 was essentially unchanged. Of course a comparison between a steroid and a protein in this respect is mis-
leading. Figure 2.3 shows the comparison between hPL and SP_1,
both proteins and likely to be subject to similar removal mechanisms.

These figures shed quite a different light on the very high
concentrations of SP_1 as compared to hPL which were recorded in
Table 2.1. These figures are deceptive. They do not reflect a
rapid production rate of the protein but its slow removal rate.
SP_1 enters the maternal plasma in a trickle, not a torrent.

Table 2.4. Plasma half-life of oestriol and placental proteins after delivery

Parameter	Half-life (min)
Unconjugated oestriol	5.8
Total oestriol	113.0
hPL	13.3
SP_1	1291.8
PAPP-A	1080.0

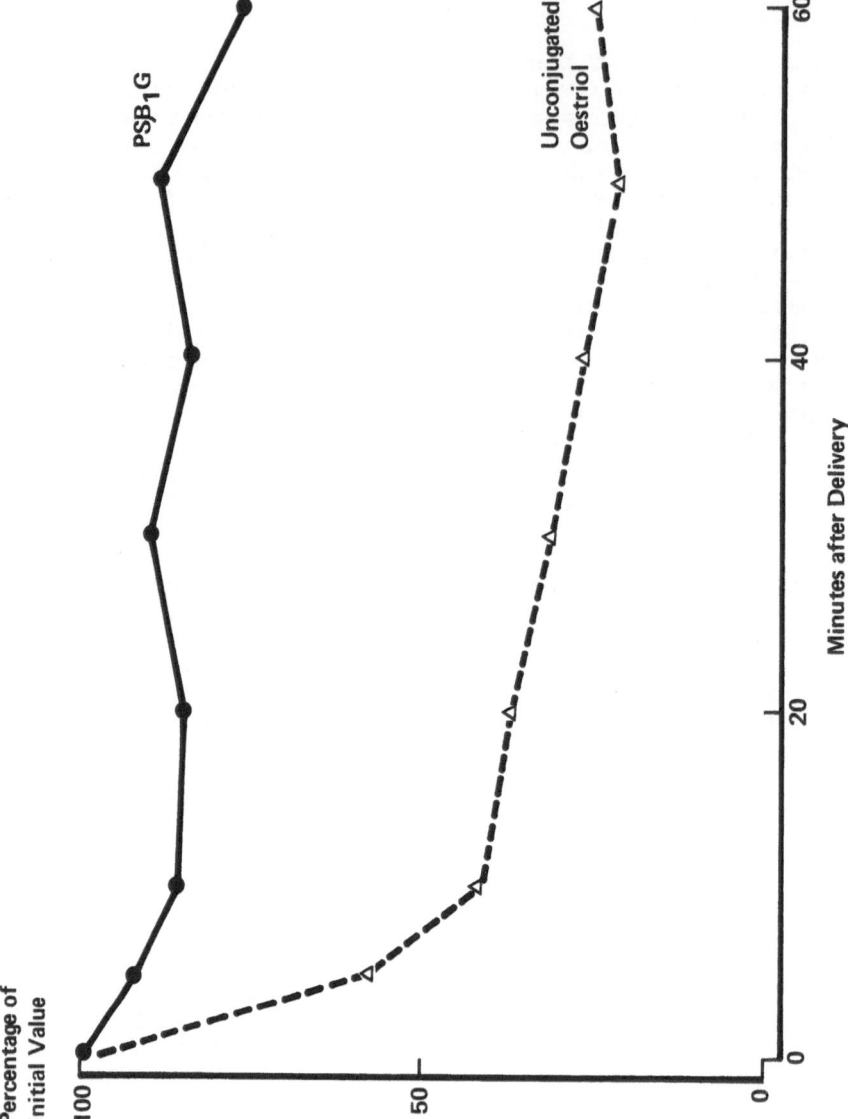

Fig. 2.2. Plasma concentraion of unconjugated oestriol and SP_1 ($PS\beta_1G$) over the first hour after delivery, relative to value at delivery

30

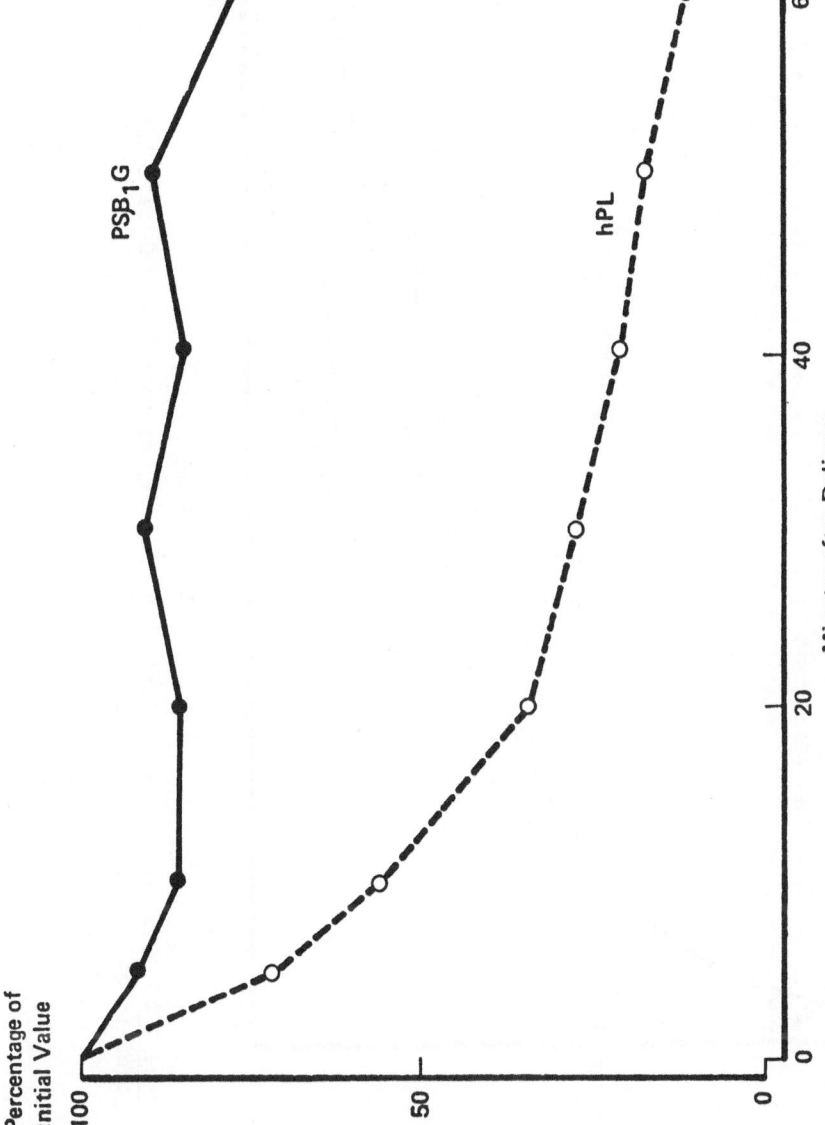

Fig.2.3. Plasma concentration of hPL and SP_1 ($PS\beta_1 G$) over the first hour after delivery, relative to the value at delivery

31

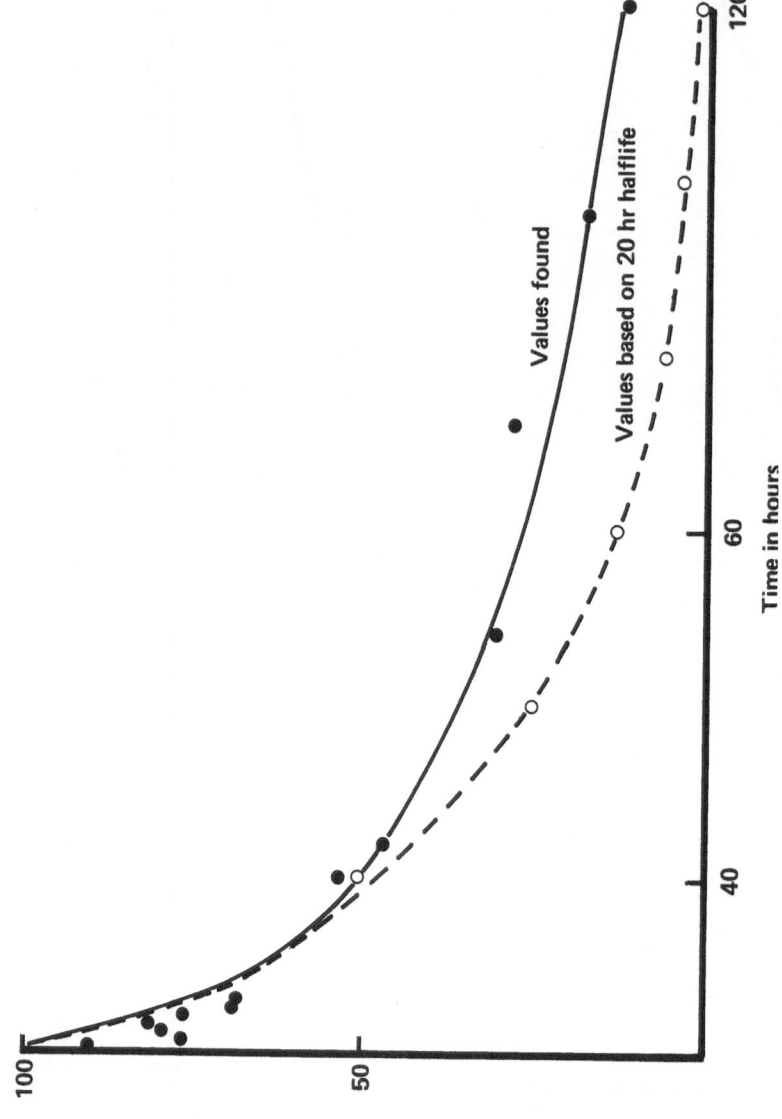

Fig. 2.4. Comparison of predicted value for SP_1 based on initial half-life as compared to values actually found

32

2.6 Compartmental Distribution

As long as there is no further inflow of protein into the maternal plasma the puerperal decline should continue at an even rate which can be predicted from the initial half-life. Figure 2.4 shows a comparison of the puerperal decline of SP_1 predicted on this basis and the values actually found. The data for hPL and PAPP-A give similar curves.

It can be seen that, although at first the values for the puerperal decline of SP_1 follow the predicted curve, after a while the rate of fall becomes noticeably slower than might be predicted on the basis of the initial half-life. The most likely explanation is that even after delivery of the placenta there is some remaining source of inflow into the maternal plasma. One possible explanation is that the proteins enter maternal compartments other than the intravascular space, and, when placental input ceases, placental protein enters the plasma from these compartments, increasingly slowing the rate of fall as the plasma concentration gets lower. Plasma proteins were, therefore, sought in other maternal compartments which communicate with the intravascular space. The relative concentrations are shown in Table 2.5.

It is evident from Table 2.5 that placental proteins leave the intravascular space to enter the interstitial fluid. Their concentration in this compartment is much less than in the plasma, but by virtue of the fact that the interstitial fluid volume is much larger than the intravascular volume, the total amount of placental protein held outside the plasma is not inconsiderable. The existence of placental protein containing spaces other than the plasma can be deduced from a different analysis of the half-life data. If the logarithm of the plasma concentration is plotted against time the values fall on a straight line as long as the emptying of only one compartment is involved. Such a log plot of the puerperal decline in hPL is shown in Fig.2.5. Similar plots can be constructed for the other proteins. This suggests that in the case of the placental proteins we are dealing with a two-compartment model.

Table 2.5. Concentration (mean and standard deviation) of oestrogens and placental proteins in various maternal compartments

Compartment	Substance				
	Oestriol nmol/litre	Oestradiol nmol/litre	SP1 mg/100 ml	PAPP-A µ/ml	hPL µ/ml
Radial Artery	77.0 ± 35.2	91.3 ± 48.9	14.7 ± 6.5	84.0 ± 45.0	5.3 ± 0.9
Peripheral Vein	61.2 ± 27.0	89.3 ± 48.0	15.6 ± 6.3	149.8 ± 117.6	5.6 ± 1.5
Peritoneal Fluid	43.6 ± 18.8	36.6 ± 21.7	4.4 ± 2.4	24.6 ± 14.1	2.3 ± 1.2

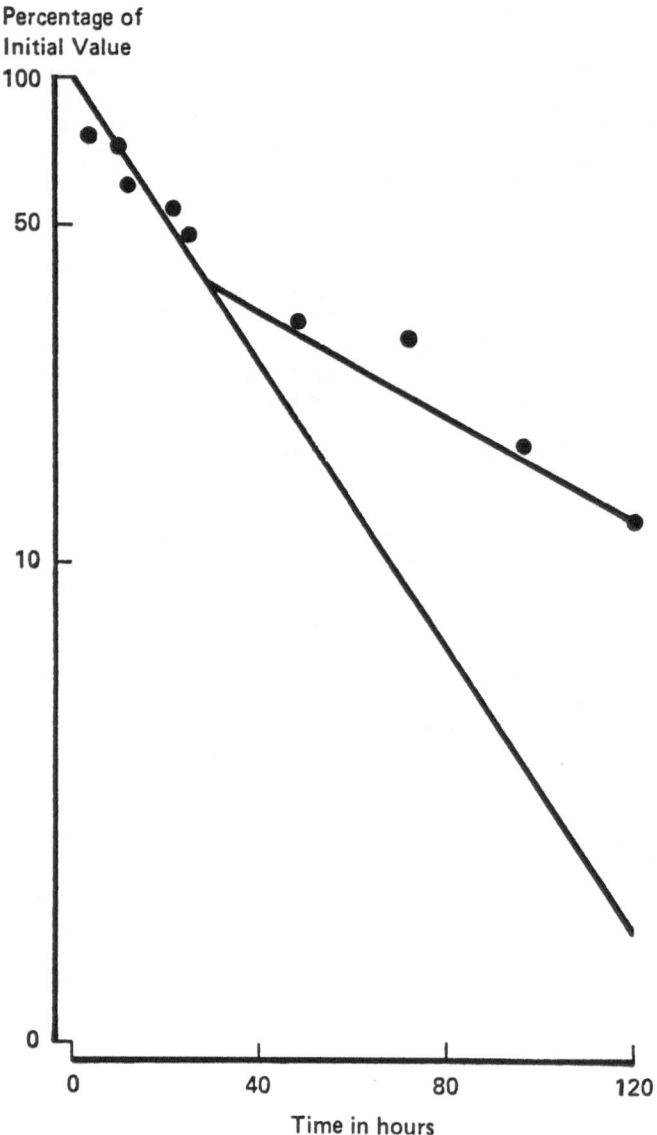

Fig. 2.5. Post-partum concentration of hPL plotted on a logarithmic scale

2.7 Route of Inflow into Maternal Plasma

Presumably a further factor which has bearing on the concentration of placental proteins in the maternal plasma is the means
by which they enter the maternal circulation. It is to be supposed that for the most part they are secreted by the syncytiotrophoblast directly into the intervillous blood. If so there
should be a gradient with high concentration in the intervillous
space, lower in the uterine vein and lowest in the peripheral
venous circulation. It is notoriously difficult to obtain representative samples of intervillous blood (FUCHS et al., 1963),

but we found we could get retroplacental blood, free from detectable contamination with fetal fluid, by sampling the blood
which welled up between the placenta and the uterine wall as
the placenta separated from the uterine wall at Caesarean section. We have accordingly collected blood from the peripheral
veins, the uterine veins and the retroplacental spaces in 15
women subjected to elective Caesarean section. In Fig. 2.6 a
comparison between the mean unconjugated oestriol in these
three spaces is shown. In order to take together the widely
different values from different individuals the uterine vein
and peripheral vein concentrations have been expressed as a
percentage of the concentration in the retroplacental blood.

When the data for Fig. 2.6 were analysed by comparing the uterine and peripheral vein values with the retroplacental oestriol
concentration in each individual subject there was no doubt of
the existence of a sharp gradient. The retroplacental value was
significantly higher than the uterine (p = 0.005) and the uterine higher than peripheral vein (p = 0.01). Of course in the
case of oestriol there is a high concentration of the steroid
in the fetal circulation, and a substantial proportion of the
retroplacental content represents, not active biosynthesis and
secretion by the trophoblast, but simple transmission by the
placenta. This does not, however, detract from the force of the
demonstration that oestriol is entering the retroplacental blood
at a high rate.

The findings for hPL, are shown in Fig. 2.7. Although less steep,
the same statistically significant gradient exists. There is

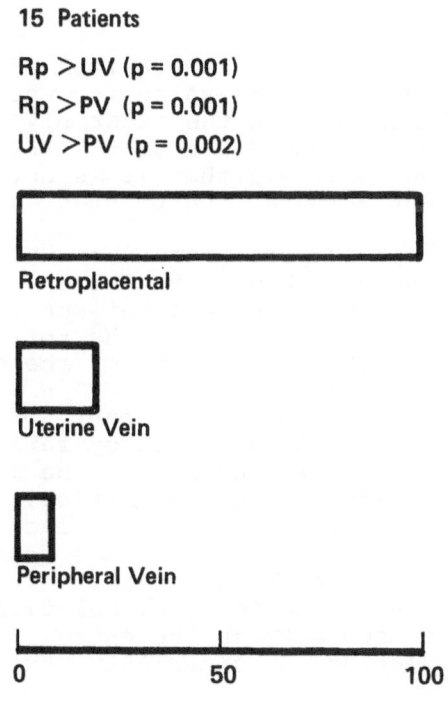

15 Patients

Rp >UV (p = 0.001)
Rp >PV (p = 0.001)
UV >PV (p = 0.002)

Retroplacental

Uterine Vein

Peripheral Vein

Fig. 2.6. Unconjugated oestriol as a
percentage of the retroplacental
value

0	50	100 %

15 Patients

Rp $>$ UV (p = 0.012)
Rp $>$ PV (p = 0.002)
UV $>$ PV (p = 0.002)

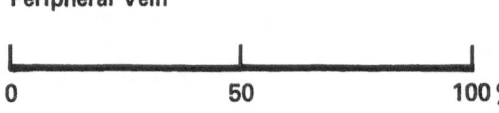

Retroplacental

Uterine Vein

Peripheral Vein

0   50   100 %

Fig. 2.7. hPL as a percentage of the retroplacental value

very little hPL in the fetal circulation and Fig. 2.7 represents, therefore, the active secretion of hPL by the trophoblast into the retroplacental space and its subsequent dispersal from this point.

The findings for PAPP-A from the same study are shown in Fig. 2.8. Although the mean value for the retroplacental blood is slightly lower than in the other sites the difference is not significant. It would appear that PAPP-A enters the maternal circulation so slowly that it is not possible to demonstrate a higher concentration at the point of entry than elsewhere.

The real surprise came when the data for SP_1 were examined. They are shown in Fig. 2.9. There is an apparent reversal of the gradient. The difference between the uterine and peripheral veins is not significant but the retroplacental blood has significantly less SP_1 than either of the other two sites ($p = 0.05$).

We are at a loss to account for this finding. A possible explanation lies in dilution by fetal blood. If there were a substantial contamination of the maternal retroplacental blood by fetal blood, this would raise, rather than lower, the oestriol concentration of the retroplacental blood as this steroid is in higher concentration in the fetal than in the maternal blood. But the same cannot be said of hPL, which exists only in trace amounts in the fetal circulation. There appears to be a real difference between the entry of hPL and of SP_1 into the maternal circulation. An alternative explanation for the lower concen-

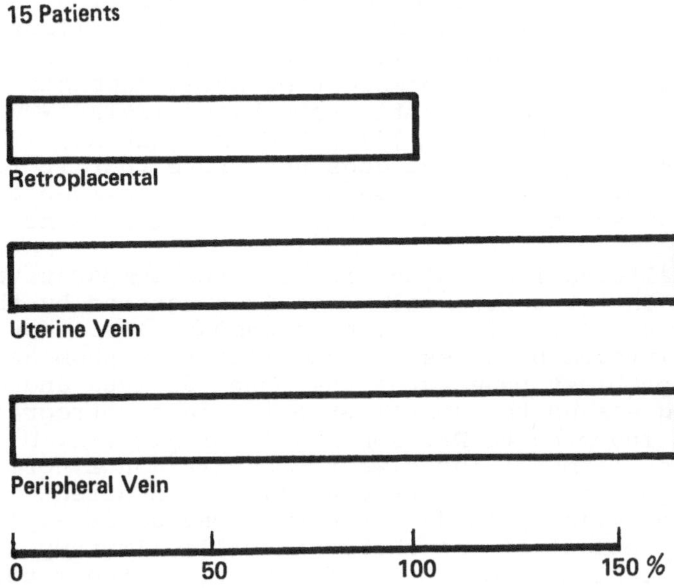

15 Patients

Retroplacental

Uterine Vein

Peripheral Vein

0	50	100	150 %

Fig. 2.8. PAPP-A as a percentage of the retro-placental value

15 Patients

RP $<$ UV (p = 0.036)
RP $<$ PV (p = 0.05)

Retroplacental

Uterine Vein

Peripheral Vein

0	50	100	150 %

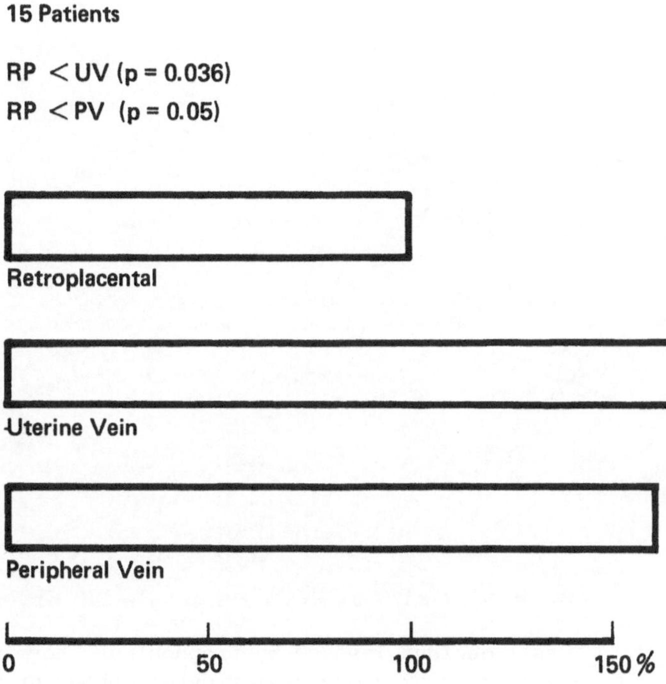

Fig. 2.9. SP$_1$ as percentage of the retropla-cental value

tration of SP$_1$ in the maternal circulation is that the chorionic villi are absorbing SP$_1$ from the intervillous blood, rather than secreting it into this space. Although a number of investigators have, by immunofluorescent methods, demonstrated that SP$_1$ is present in the syncytiotrophoblast (TATARINOV et al., 1976; BOHN, 1972b; HORNE et al., 1976; LIN and HALBERT, 1976), the demonstration of its presence does not prove that it is made there. There is, nevertheless, so much tangential evidence that the trophoblast synthesizes SP$_1$ that this belief is not being challenged.

An alternative explanation for the comparatively low level of SP$_1$ in retroplacental blood is suggested by the work of ROBERTSON et al., 1976. They have demonstrated that the uterine wall is invaded by waves of non-villous trophoblast. Some invasive trophoblast penetrates the basal decidua and other portions progress up the spiral arteries in a retrograde direction. We are indebted to Dr. Robertson for Figs. 2.10 and 2.11 which show these forms of invasive trophoblast in a normal pregnancy. These trophoblast cells are clearly not in direct contact with the intervillous blood in the same way as it applies to the syncytiotrophoblast of the chorionic villi. If they are a source of a substantial proportion of the SP$_1$ in the maternal circulation, this protein is likely to make its way into the maternal circulation via channels of uterine drainage not involved with the retroplacental space.

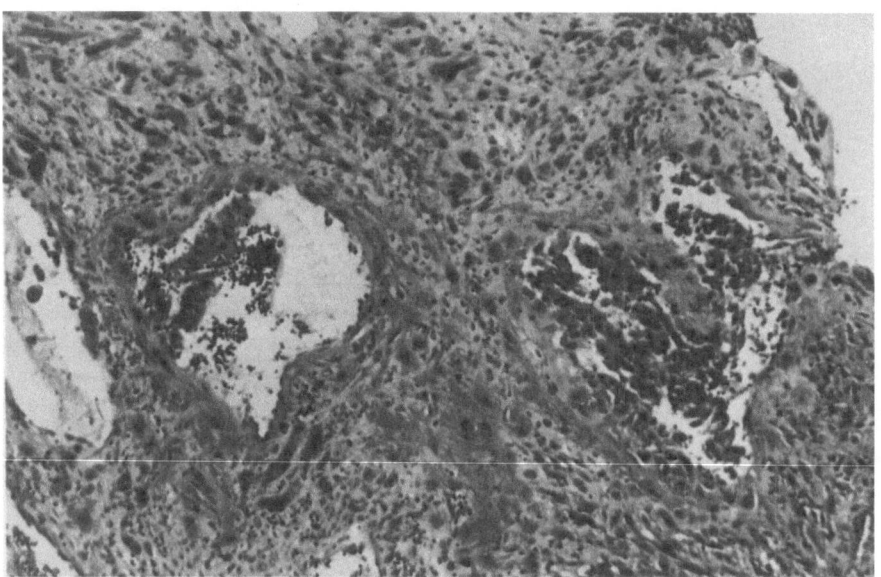

Fig. 2.10. Basal decidua invaded by trophoblast, termination of maternal spiral artery plugged by endovascular cytotrophoblast which is also partially incorporated into the fibrinoid vessel wall x 80. Reproduced by permission from *Hypertension in pregnancy*. LINDHEIMER, M.D., KATZ, A.L., ZUSPAN, F.P. (eds.). New York: John Wiley & Sons 1976

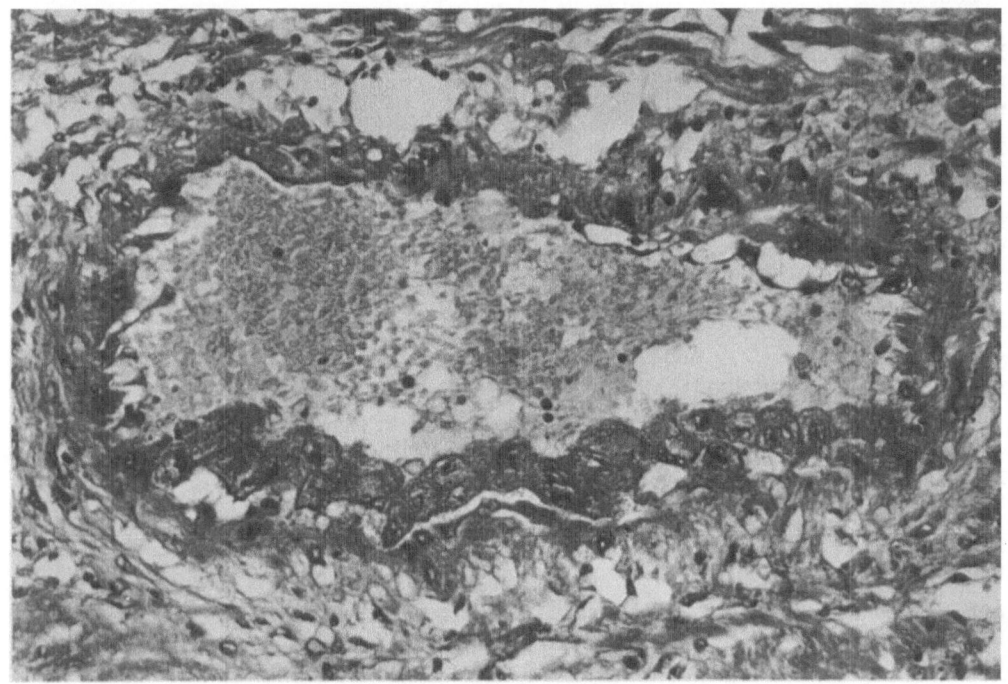

Fig. 2.11. Interstitial cyto- and syncytiotrophoblast in basal decidua; endovascular cytotrophoblast in a decidual artery with early physiological changes. Reproduced with permission from *The pathology of pregnancy*. HURLEY, R. (ed.). London: Royal College of Pathologists 1976

This suggestion puts the SP_1 in the maternal plasma in a different light. It is not, like hPL, a hormone which is being secreted by the placenta into the maternal circulation in order to bring about some metabolic effect in the mother. It is a locally active agent, performing some function within the uterine tissue, and the material found in the maternal peripheral circulation is waste, leaking away from the site of action, on its way to disposal. If this concept is valid it detracts from the likely usefulness of assays of SP_1 in peripheral blood as a means of assessing placental function but lays stress on its functional role in pregnancy. It could well be part of the mechanism which prevents rejection of the conceptus, but in the sense of local tissue activity rather than of a suppressive action on the central immune systems of the mother.

References

Bohn, H.: Nachweis und Charakterisierung von Schwangerschaftsproteinen in der menschlichen Plazenta, sowie ihre quantitative immunologische Bestimmung im Serum schwangerer Frauen. Arch. Gynäkol. 210, 440-457 (1971)

Bohn, H.: Isolierung und Charakterisierung des schwangerschafts-spezifischen β_1-Glykoproteins. Blut 24, 292-302 (1972a)

Bohn, H.: Nachweis und Charakterisierung von löslichen Antigenen in der menschlichen Plazenta. Arch. Gynäkol. 212, 165-175 (1972b)

Bruce, D., Klopper, A.: The measurement of pregnancy-specific β_1 Glycoprotein by electroimmunodiffusion. Clin. Chim. Acta 84, 107-113 (1978)

Fuchs, F., Spackman, T., Assali, N.S.: Complexity of the intervillous space. Am. J. Obstet. Gynecol. 86, 226-231 (1963)

Gall, S.A., Halbert, S.P.: Antigenic constituents in pregnancy plasma which are undetectable in normal, non-pregnant female or male plasma. Int. Arch. Aller. Appl. Immunol. 42, 503-515 (1972)

Horne, C.H., Towler, C.M., Pugh-Humphreys, R.P., Thompson, A.W., Bohn, H.: Pregnancy-specific β_1 glycoprotein, - a product of the syncytiotrophoblast. Experientia 32, 1197 (1976)

Klopper, A., Masson, G., Wilson, G.: Plasma estriol and placental proteins: A cross-sectional study at 38 weeks gestation. Brit. J. Obstet. Gynaecol. 84, 648-655 (1977)

Klopper, A., Buchan, P., Wilson, G.: The puerperal decline of oestriol and pregnancy associated proteins. Brit. J. Obstet. Gynaecol. (1978) (in press)

Lin, T.M., Halbert, S.: Placental localisation of human pregnancy-associated plasma proteins. Science 193, 1249-1252 (1976a)

Lin, T.M., Halbert, S.P., Kiefer, D., Spellacy, W.N., Gall, S.: Characterization of four pregnancy-associated plasma proteins. Amer. J. Obstet. Gynecol. 118, 223-236 (1974)

Lin, T.M., Halbert, S.P., Spellacy, W.N.: Relation of obstetric parameters to the concentration of four pregnancy-associated plasma proteins at term in normal gestation. Am. J. Obstet. Gynecol. 125, 17-24 (1976b)

Lindberg, B.S., Nilsson, B.A.: HPL levels in normal pregnancy. J. Obstet. Gynaecol. Brit. Cwlth. 80, 619 (1973)

Masson, G., Klopper, A., Wilson, G.: Plasma estrogens and pregnancy associated plasma proteins: A study of their variability in late pregnancy. Obstet. Gynecol. 50, 435-438 (1977)

Robertson, W.B., Brosens, I., Dixon, G.: Uteroplacental vascular pathology. Eur. J. Obstet. Gynecol. Reprod. Biol. 5, 47-65 (1976)

Tatarinov, Y.S. Masyukevich, V.N.: Immunochemical identification of new beta$_1$ globulin in the blood serum of pregnant women. Bull. Eksp. Biol. Med. 69, 66-68 (1970)

Tatarinov, Y.S., Falaleeva, D.M., Kalashnikov, V.V., Tatra, G., Breitenecker, G., Gruber, W.: Serum concentration of pregnancy-specific β-1-glycoprotein (SP-1) in normal and pathological pregnancies. Arch. Gynäk. 217, 383 (1974)

Toloknov, B.O.: Immunofluorescent localisation of human pregnancy-specific β-globulin in placenta and chorioepithelioma. Nature 260, 263 (1976)

Towler, C.M., Horne, C.H., Jandial, V., Campbell, D.M., MacGillivray, I.:
  Plasma levels of pregnancy specific B_1-glycoprotein in normal pregnancy.
  Brit. J. Obstet. Gynaecol. $\underline{83}$, 775 (1976)

3 Potential Antifertility Vaccines Using Antigens of hCG

V. C. Stevens

Despite the knowledge of the existence of human chorionic gonadotrophin (hCG) for nearly 50 years, the functions of this hormone during pregnancy are still incompletely understood. The
hormone has been detected in the maternal circulation as early
as 8 days following conception and recently, indirect evidence
has been provided to suggest that trophoblast cells of the pre-
implanted blastocyst synthesize hCG. Endocrinologists have believed for many years that the principal role of hCG in early
pregnancy is to "rescue" the corpus luteum from its normal regression following ovulation and to stimulate this structure to
secrete ovarian steroids for uterine maintenance until placental
steroidogenesis is initiated. Also, many immunologists have suggested various roles for hCG in preventing rejection of the conceptus existing normally in the maternal environment as a foreign
allograft. At this time, the precise role of the hormone during
normal gestation has not been defined.

Antibodies to hCG, raised in a variety of species, have been used
extensively for studies of reproductive biology. The well-docu-
mented cross-reactivity of such antibodies with human luteinizing hormone (hLH) have permitted their use in immunoassay for
both hormones. The ability of hCG antisera to neutralise the
biological activity of both hormones has been observed and this
property used to study mechanisms of gonadotrophin action. Some
antisera to hCG have been shown to cross-react with LH or CG of
nonhuman primates and sera have been used to demonstrate the importance of gonadotrophins for corpus luteum function. These

observations provided a basis for the concept that the induction of antibodies to hCG in women may provide an effective means of fertility regulation.

From the outset, it was recognized that development of a method to regulate human fertility by immunisation against hCG would be possible only after certain fundamental problems were overcome. While this task is by no means simple, the problems anticipated can be categorized into two general areas: immunogenicity and specificity. Since hCG is normally well tolerated immunologically by women during pregnancy and in both men and women following treatment with the hormone for infertility, means must be devised to render hormone isoantigens immunogenic in humans. Further, preparations must elicit an immune response in women, capable of inhibiting fertility without the use of noxious adjuvants, such as Freund's complete, which are unacceptable for human use. The second major problem, nonspecificity of hCG antibodies, must also be overcome. The well-known cross-reactivity of hCG antisera with hLH provides a serious threat to the safety of any method due to potential autoimmune damage to the maternal pituitary or to potential immune diseases such as nephritis. With full realization of the magnitude of the obstacles for development of such a method, studies were initiated more than a decade ago to investigate the feasibility of an immunological method for fertility regulation based on immunisation against hCG.

3.1 Studies to Overcome Natural Tolerance to Gonadotrophins

Since nonhuman CG preparations were not available and since hCG and hLH are biologically and immunologically very similar, initial studies were performed in baboons using pituitary hormone preparations. Model studies designed to test hypotheses on immunological tolerance, such as those reported by CINADER and DUBERT (1955) provided experimental procedures for these studies. Their report demonstrated that chemical alteration of tolerated antigens by hapten-coupling could provide immunogens capable of eliciting antibodies to unaltered antigens. These procedures were applied to the problem of rendering homologous gonadotrophins immunogenic in baboons (STEVENS, 1973). These studies revealed that hapten-coupling of baboon LH (bLH) with diazotized sulphanilic acid provided a preparation which, when used to immunise female baboons, elicited antibodies reactive to native bLH and disrupted ovulation and ovarian steroid secretion as long as significant antibody levels persisted. It was also observed that return of endogenous LH secretion did not evoke a secondary rise in antibody production and immunised animals were fertile when the immune response had waned. Antifertility effects could, however, be reestablished by a "booster" injection of hapten-coupled bLH. These studies encouraged the belief that chemical alteration of gonadotrophins could provide a means to overcome natural tolerance to hormones following isoimmunisation.

Additional studies were conducted to address the specific question, "Can hCG be altered to render it immunogenic in humans?" Hapten-coupled hCG was administered to postmenopausal and surgically sterilised women and antibody production and hormone patterns assessed (STEVENS and CRYSTLE, 1973). While the older women showed only a weak antibody response to hCG, the younger women (aged 24-35) produced higher antibody levels, and normal patterns of LH and ovarian steroid secretion were disrupted.The antibodies generated by these women cross-reacted strongly with hLH and no reimmunisation was performed.

While these studies did not suggest a method that could be applied widely to human populations, they did provide some credibility to the basic concept of inhibiting endogenous hormone function by isoimmunisation with chemically modified antigens. Having demonstrated that one of the major problems in developing an hCG vaccine (immunogenicity) could be overcome under certain conditions, efforts were begun to identify an hCG antigen that could be used to produce antibodies specific for hCG with little or no reaction with hLH.

3.2 Studies Using the β Subunit of hCG as a Potential Vaccine Antigen

A stimulus to this research effort was provided by the report by VAITUKAITIS et al. (1972) that antibodies raised to the hCG β subunit (β-hCG) could discriminate between hCG and hLH sufficiently well to be used in a specific radioimmunoassay for hCG. There was cross-reactivity with hLH in their assay at high doses but the β-hCG preparation used to generate antisera was not totally free of intact hCG and, therefore, the cross-reaction could be attributed to some antibodies to the whole hormone. Considering that it may be possible to produce completely hCG-specific antibodies with "pure" β-hCG, studies were initiated to determine whether this subunit would provide a suitable antigen for use in a fertility-regulating vaccine.

A preparation of β-hCG was isolated from highly purified hCG in which the final step was an affinity chromatography procedure using an antiserum to hCG α subunit. This method provided an hCG-β preparation with very low contamination with intact hCG as evidenced by a biological activity of less than 1.0 IU/mg. This preparation was used to immunise rabbits, sheep, cows and baboons. Antisera were evaluated for specificity by studying their ability to bind antigens in vitro and to neutralise hormone activity in vivo in rodents (STEVENS, 1976). All antisera produced cross-reacted with hLH. Some antisera, produced in rabbits, reacted relatively little to hLH but in the majority of animals the cross-reactivity was greater that 15%.

Antisera raised in baboons reacted with hLH, baboon CG (bCG) but not bLH. Immunised females with relatively high antibody levels showed normal patterns of hormone secretion during the cycle and ovulation occurred regularly. However, when these females were mated repeatedly with males of proven fertility, no animal failed to have menses at or before the expected time of the cycle. The stage of gestation disrupted by immunisation was not apparent but no significant change in hormone patterns from the normal cycle was detected, suggesting a preimplantation or immediate post-implantation disruption. Thus the efficacy of CG neutralisation for fertility inhibition in a primate species was demonstrated.

Despite the additional support these data gave to the idea of an anti-hCG vaccine, the significant cross-reactivity of anti-β-hCG sera with hLH indicated that human immunisation with this sub-unit would be unlikely to produce specific immune responses to hCG. While some held the view that β-hCG would be a safe immunogen for humans (TALWAR et al., 1976), the author and co-workers decided that there were too many theoretical hazards to non-specific immunisations and sought alternative, more specific antigens for method development.

3.3 Studies Using Peptide Fragments of β-hCG as Potential Vaccine Antigens

The amino acid sequences of both hLH and hCG were published in 1973 (MORGAN et al., 1973). The α subunits of the two hormones have identical structures. Examination of the β subunits of hCG and hLH reveals that hCG contains 145 amino acid and hLH only 115. Only 16 of the residues between positions 1 and 110 are different between the subunits, and the remaining 94 residues (85%) of this portion of the molecule are the same for both hormones. Therefore, it is not difficult to understand why antisera raised to β-hCG readily cross-react with hLH. However, the C-terminal 35 residues of β-hCG (residues 111-145) share only one residue in common with hLH (Asp. at 112). The availability of this structural information offered further opportunities to evaluate antigens that might elicit specific antibodies to hCG.

Studies were initiated to isolate C-terminal fragments of β-hCG and to test the specificity of antibodies raised to them. At the same time, synthetic peptides representing the same amino acid sequence as the natural peptide were prepared for testing. Since the natural peptides contained four separate carbohydrate chains, it was important to determine whether synthetic peptides would provide antigens suitable for generating hCG-specific antibodies. This question was critical to the use of C-terminal β-hCG fragments for a vaccine, since the production of sufficient quantities of peptides from natural sources would not be feasible.

Antisera generated to both natural and synthetic peptides representing residues 109-145 of β-hCG reacted similarly to intact hCG. Antibodies to the two preparations were somewhat different in their reactivities to natural and synthetic peptides; however, the use of synthetic peptides for production of hCG antisera was shown to be feasible (Fig. 3.1). Also, when injected under identical conditions, the natural peptide was not more immunogenic than the synthetic one. More important, antisera to neither natural nor synthetic peptides showed detectable cross-reaction with hLH. Peptides representing 35 or more residues of the β-hCG C-terminus were capable of evoking antibodies which would neutralise the biological effect of hCG in vivo. These

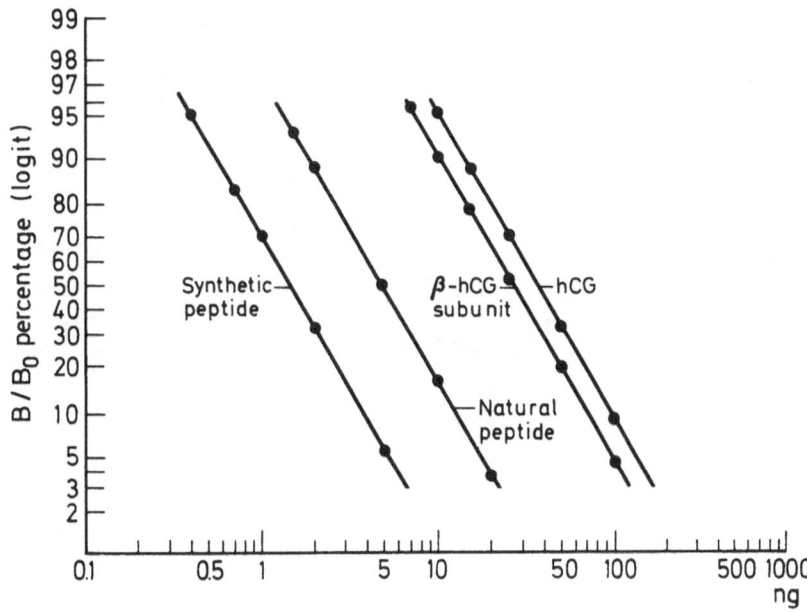

Fig. 3.1. Reactivity of C-terminal peptides, β subunit, and intact molecules of hCG to an antiserum raised to a synthetic C-terminal β-hCG peptide. Dose-response curves represent displacement of ^{125}I-labelled synthetic peptide from antibodies by unlabelled antigens. B, amount bound in presence of competitor; Bo, amount bound in absence of competitor. (Reprinted from: Physiological effects of immunity against reproductive hormones. EDWARDS, R.G., JOHNSON, M.H. (eds.), p. 261. Cambridge: Cambridge University Press 1976)

findings suggested that the use of a 35-residue synthetic peptide might be a suitable antigen for vaccine development.

The enthusiam for identifying an antigen capable of eliciting neutralising antibodies to hCG without any detectable reaction with hLH was somewhat dampened by the observation that sera from non-pregnant women, men and several laboratory animals contained a substance reactive to antipeptide sera. When radioimmunoassays employing antisera to peptides and I^{125} hCG were used to assay serum or urinary extracts, positive responses were always obtained even though most samples were from subjects who were not believed to be producing any hCG. Further data were reported by CHEN et al. (1976) indicating that positive responses were obtained with human pituitary extracts using a similar antipeptide serum radioimmunoassay. These findings were confirmed in the author's laboratory and numerous side fractions obtained from the isolation of pituitary hormones have shown varying degrees of activity in these assay systems (Fig.3.2). The level of hCG-like material in these extracts is very low and their relative potency is about 1000 times less than highly purified hCG. Also, hCG-like material has been detected in tissue extracts of ovary, kidney and intestine using assays employing antisera to β-hCG. Some workers have suggested that these

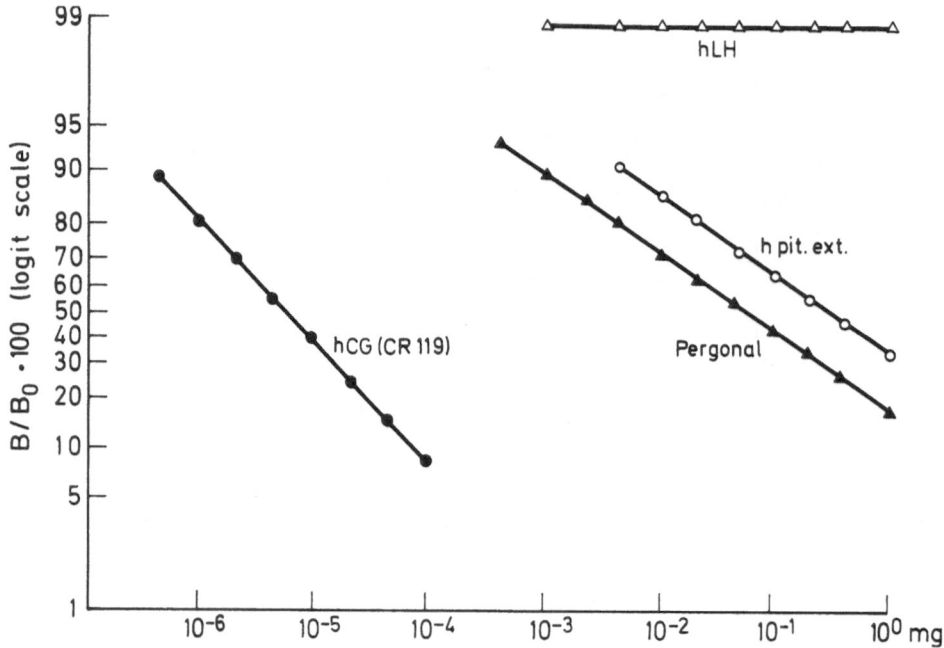

__Fig. 3.2.__ Reactivity of pituitary and urinary (Pergonal) extracts to anti-serum raised to a synthetic C-terminal β-hCG peptide. The radioimmunoassay employed ^{125}I hCG and an antiserum to C-terminal β-hCG peptide 111-145. Responses are plotted the same as in Fig. 2.1

responses may be due to bona fide hCG produced in small amounts by normal tissues. These findings are confusing and worrying if it is proposed to use peptide antigens in a human vaccine. On the other hand, some of the same antisera that were used in assays showing positive values for pituitary extracts, failed to react with sections of anterior pituitaries using an immuno-fluorescent technique. Female baboons actively immunised to these peptides, have not shown any detectable alteration of hormone levels or other manifestation of the presence of antibodies. At this time, neither the question of the specificity of antibodies to C-terminal β-hCG peptides nor the existence of hCG in normal tissues has been answered.

In addition to the problem discussed above with vaccines using C-terminal β-hCG peptides, the immunogenicity of these peptides is not adequate to induce high levels of antibodies to hCG. In this case, it is not primarily a problem of breaking natural tolerance to the antigen but one of molecular size. Peptides with molecular weights of 3000-4000 are weakly antigenic even when administered with Freund's complete adjuvant in a heterologous species such as rabbits. The usual method of generating antibodies to small molecules by conjugating them to macromolecular proteins has been applied to these hCG peptides in an effort to provide an immunogen capable of eliciting adequate immune responses. Proteins used as carriers in these conjugates were selected for their suitability in humans. Most experiments have been performed utilizing tetanus toxoid as a carrier. While there are numerous procedures reported in the literature for chemical conjugation of small molecules to carriers, most studies were performed to obtain antibodies for qualitative characterization or for use in immunoassays. No reports were found that were specifically directed towards providing immugens for therapeutic application. Coupling peptides or haptens to proteins was usually accomplished by use of carbodiimide, glutaraldehyde, diisocyanates and other cross-linking reagents. The use of these reagents to couple hCG peptides to carriers resulted in conjugates eliciting only low levels of antibodies in mice, rabbits and baboons. Examination of these conjugates suggested that the reason for these weak responses may have been either a) too few peptides attached to each carrier molecule or b) polymerisation of peptides before conjugation which did not permit exposure of the complete peptide sequence to the immune system. Intensive research has been initiated to devise new methods of conjugation to overcome these problems and currently several different conjugates, prepared by a variety of chemical coupling methods, are being evaluated for immunogenicity.

Despite the low levels of antibodies obtained to the initial conjugates prepared, female baboons immunised with these materials were mated with males of proven fertility. Somewhat surprisingly, a significant reduction in fertility was observed in the few animals tested. Four females immunised with one conjugate were mated three times (12 exposures) with only one conception (Table 3.1). Since the fertility rate in control animals was approximately 85%, there was little doubt that an antifertility effect had been induced. In addition, it was found that the antibodies produced reacted with baboon CG only about 10% as well as with hCG. The conclusion reached from these results was either (a) very low levels of CG antibodies are able to block fertility or (b) cell-mediated immune responses (CMI) played a role in fertility disruption following immunisation.

Studies have been performed to ascertain whether humoral immune responses (antibodies) or CMI responses (lymphocytes or macrophages) were the more important for the antifertility effect observed. The effects of antibodies were easier to study, since CMI responses must be tested separately for each conjugate prepared. Antisera were pooled from a number of baboons immunised with the β-hCG peptide 111-145 and 100 ml was administered intravenously to a nonimmunised pregnant baboon in early gestation. An abortion occurred within 36 h. Since implantation was well established at the time of passive immunisation, the results suggested that CG from the baboon was neutralised and/or the antibodies exerted a cytotoxic action on the placenta. Further study will be required to define precisely the actions of such antisera.

CMI responses to conjugates have not yet been completely characterized. Blastogenic transformation of peripheral lymphocytes of immunised baboons in vitro has been induced by additions of hCG, peptides and carriers to the cell-containing media.

Table 3.1. Breeding of baboons immunized with a synthetic peptide-protein conjugate

| Baboon Nr. | Mating 1[a] | | | Mating 2 | | | Mating 3 | | |
	Pre-mating[b] titer (ng/ml)	Ovulated	Pregnant	Pre-mating[b] titer (ng/ml)	Ovulated	Pregnant	Pre-mating[b] titer (ng/ml)	Ovulated	Pregnant
1374	108	+	–	110	+	–	97	+	–
1422	123	+	–	101	+	–	100	+	–
976	345	+	–	331	+	–	305	+	–
1068	95	+	–	89	+	–	91	+	+

[a] Mating 1 commenced at approximately 3 months after immunization.
[b] Binding to ^{125}hCG/ml serum.

The type of lymphocytes stimulated in the in vitro tests has not been defined although they are presumed to be mostly T-lymphocytes. Positive baboon skin tests to carrier proteins only have been observed. Further studies are in progress to characterise CMI responses to these peptide conjugates.

The success of current efforts to develop an hCG vaccine will depend upon whether the available peptide-carrier conjugates will inhibit fertility in nonhuman primates and whether the immune responses to these immunisations can be shown to be without unacceptable side-effects or health hazards. At the time of writing, a large number of animals have been immunised in an attempt to answer these questions. In the paragraphs below, an outline of the studies thought to be essential to method evaluation is presented.

3.4 Studies Necessary to Evaluate Efficacy and Safety of a hCG-Peptide Vaccine

The development of an anti-hCG vaccine for regulating human reproduction constitutes a new concept to therapeutics. No precedent of an immunological method for reducing fertility is available for comparison. Further, no vaccine in current use employs a synthetic component or a peptide-carrier conjugate. These and other unknown factors make it mandatory that a careful and thorough evaluation of any vaccine be made in animals prior to human trials.

Animal models selected for vaccine assessment are critical to meaningful results. First of all, the antibodies and/or CMI responses elicited to the human antigen must react to some degree with endogenous CG in the model species. In this regard, anti-hCG peptide sera react well with gorilla and chimpanzee CG, rather weakly with baboon CG and very weakly with Rhesus monkey CG. Marmosets have been used successfully to demonstrate antifertility effects of antisera to β-hCG (HEARN, 1976) although the immunological similarity of CG from this species and hCG is not yet well defined.

The animals used for these studies must have a high fertility rate and preferably, remain fertile throughout the year. For this and other reasons, baboons and chimpanzees have been selected for evaluating the efficacy and safety of potential hCG peptide vaccines. The bulk of the assessment will be performed in baboons, where sacrifices for pathological studies are possible, and studies in chimpanzees will be limited to assessment of the functional integrity of various physiological systems in immunised females.

The principal test of efficacy of the method is, of course, the pregnancy rate of immunised females after repeated matings with males of proven fertility. Characterization of the immune response is also important to determine the mechanism of fertility inhibition and to observe the variation in immune responsiveness among individuals. Outbred populations of baboons and humans are likely to exhibit substantial variation in immune responsiveness due to genetic factors and the frequency of "non-responders" must be ascertained. The acceptability of a method will require this number to be very small.

The safety of an hCG vaccine will primarily depend upon the specificity of the immune response. Antibody and/or lymphocyte attack on any tissue other than the conceptus could result in serious health hazards and would clearly contraindicate the use of the procedure. Complete evaluations of hormone levels and functions in immunised, non-pregnant females must be performed. Reaction of antisera to any normal human or animal tissue in vitro should be followed by a careful assessment of the function of that tissue in vivo. Currently, studies are in progress to determine whether the observed weak reaction of anti-peptide antibodies with pituitary and urinary extracts would pose any hazard to immunised individuals.

A theoretical potential hazard to an hCG vaccine is the induction of immune complex diseases. Should the antibodies in the circulation react with secreted hCG shortly after conception to form complexes of a certain size, these could lodge on the membrane of the glomerulus of the kidney and possibly result in nephritis. This condition could also result if antibodies cross-reacted with normal body components (such as pituitary LH). Therefore, for adequate safety evaluation, it is essential that the antigen selected should elicit antibodies cross-reacting with the same non-CG components in the model species as they do with human tissues or secretions.

The reversibility of the vaccine should be tested. Whether established immunity will wane after a predictable period and render the female fertile again will determine whether the method can be used only for sterilisation or as a child-spacing technique. Also, teratological effects on fetuses from method failures or immunisations during an undetected pregnancy must be studied. The incidence and severity of immediate or delayed hypersensitivity reactions will also determine whether unacceptable risks are associated with the vaccine.

The evaluations outlined above by no means include all of the tests and studies that must be completed before an hCG vaccine could be applied to human use. However, the major areas are included here to illustrate the complexity of developing an immunological procedure for human fertility regulation. Despite these problems and obstacles, the potential advantages offered by such a method warrant continuous and careful work towards this goal.

References

Chen, H., Hodgen, G., Matsura, S., Lin, L.J., Gross, E., Reichert, L.E., Birken, S., Canfield, R.E., Ross, G.T.: Evidence for a gonadotrophin from nonpregnant subjects that has physical, immunological, and biological similarities to human chorionic gonadotrophin. Proc. Natl. Acad. Sci. U.S.A. 73, 2885-2889 (1976)

Cinader, B., Dubert, J.M.: Acquired immune tolerance to human albumin and the responses to subsequent injections of diazo human albumin. Brit. J. Exp. Pathol. 36, 515-529 (1955)

Hearn, J.P.: Immunisation against pregnancy. Proc. R. Soc. Lond. Biol. 195, 149-160 (1976)

Morgan, F.J., Birken, S., Canfield, R.E.: Human chorionic gonadotrophin. A proposal for the amino acid sequence. Mol. Cell. Biochem. 2, 97-99 (1973)

Stevens, V.C.: Immunisation of female baboons with the hapten-coupled gonadotrophins. Obstet. Gynecol. 42, 496-506 (1973)

Stevens, V.C.: Actions of antisera of hCG-β: *in vitro* and *in vivo* assessment. In: Proceedings of the international congress of endocrinology. JAMES, V.H.T. (ed.), pp. 379-385. Amsterdam: Excerpta Medica 1976

Stevens, V.C., Crystle, C.D.: Effects of immunisation with hapten-coupled hCG on the human menstrual cycle. Obstet. Gynecol. 42, 485-495 (1973)

Talwar, G.P., Sharma, N.C., Dubey, S.K., Salahuddin, M., Des. C., Ramakrishnan, S., Kumar, S., Hingorani, V.: Isoimmunisation against human chorionic gonadotrophin with conjugates of processed β-subunit of the hormone and tetanus toxoid. Proc. Natl. Acad. Sci. U.S.A. 73, 218-222 (1976)

Vaitukaitis, J.L., Braunstein, G.D., Ross, G.T.: Radioimmunoassay which specifically measures human chorionic gonadotrophin in the presence of human luteinizing hormone. Am. J. Obstet. Gynecol. 113, 731-758 (1972)

4 The Use of Antibody Affinity Chromatography and Other Methods in the Study of Pregnancy-Associated Proteins

R.G.Sutcliffe, B.M.Kukulska, L.V.B.Nicholson, and W.F.Paterson

The search for new pregnancy associated proteins has involved many disciplines and included many enzymatic, hormonal, electrophoretic and immunological techniques (see BOHN, 1976). Enzymatic and hormonal methods are usually specific for proteins of known function or effect, whose nature must, therefore, be defined before the protein can be detected. By contrast, electrophoretic and immunological procedures can be carried out without assumption as to the function of the protein being sought. The broad applicability of electrophoretic and immunological methods is, however, offset by their limited ability to distinguish tissue-specific proteins which exist in low concentration amongst the very wide array of intracellular and extracellular proteins found in virtually all tissues. Thus, immunisation of rabbits with 100-300 mg of protein from extracts of human placenta or human fetal liver yielded precipitating antibodies against the proteins commonly found in adult human serum. After adsorption with solid phase adult human serum protein (AVRAMEAS and TERNYNCK, 1969), only some ten precipitin arcs could be detected by antibody-antigen crossed electrophoresis (AACE). We have used two approaches to reduce the strong antibody response to high concentration serum proteins. The first

derives from the work of GOLD and FREEDMAN (1965), who raised
rabbit antibodies against human carcinoembryonic antigen by in-
ducing neonatal tolerance in the rabbits to extracts of normal
human colon before priming the rabbits with extracts of colonic
tumour. The second concerns the removal of adult-type serum proteins from fetal material by use of large scale immuno-absorbent
affinity columns using antibodies raised against adult human
serum proteins.

4.1 The Use of Neonatal Immune Tolerance

Four newborn rabbits were injected subcutaneously with 80 mg
adult human serum protein and 40 mg soluble protein from adult
human red cell lysates, between the first and seventieth day of
life (GOLD and FREEDMAN, 1965). In a separate series of experiments a further four neonatal rabbits were injected with 100 mg
soluble protein from adult human liver. After 80 to 90 days all
animals were challenged with 60 mg soluble protein from extracts
of human fetal liver (material kindly provided by Dr.A.A.M. Gib-
son and by Dr. D.J.H. Brock), suspended as a dry powder in
Freund's complete adjuvant. Booster injections were given at
monthly intervals and the antibody response in each rabbit was
monitored by AACE.

At the end of the course of tolerisation, the rabbits tolerised
with adult human serum and red cell lysate were producing precipitating antibodies to between three and five adult human
serum proteins. Rabbits tolerised with protein from adult human
liver homogenates showed almost complete tolerance to human serum proteins. Subsequent injection of fetal liver in adjuvant
slightly intensified any serum protein precipitin arcs.

Antisera from rabbits tolerised with adult blood or adult liver
were adsorbed with solid phase adult human serum and were tested
for residual specificities on AACE using protein from homogenates of fetal and adult liver as test antigens. Rabbits tolerised with adult liver and then immunised with fetal liver developed no antibodies which were detectable on AACE against fetal liver. Thus, no putative fetal-specific liver proteins were
detected with this protocol.

Four rabbits which had been tolerised with adult human serum and
red cell lysate and three untreated rabbits were immunised with
60 mg fetal liver protein. The antisera were absorbed with adult
human serum and were then tested by AACE against protein from
adult and fetal liver. The results are shown in Table 4.1.

The number of precipitin arcs was similar in both adult and fetal liver, and differed little between the control and tolerised
groups. All animals produced antibodies to ferritin. After adsorption with solid phase adult liver protein, no fetal-specific
antigens were found in fetal liver.

These results suggest that there is no great advantage in tolerising rabbits to adult protein before immunising with extracts

Table 4.1. Responses to immunisation with 60 mg human fetal liver. The figures refer to the number of precipitin arcs observed when antisera against human fetal liver were raised in rabbits, a group of which had been tolerised with human serum and red cell proteins. The antisera were adsorbed with adult serum and tested on AACE using extracts of adult liver (*AL*) and fetal liver (*FL*) as test antigen. The control rabbits were studied in the same way, except that they received no tolerising antigen during the neonatal period

Rabbit	Booster number					
	One		Two		Three	
Controls						
6989	6	5	6	6	NT[a]	NT
6990	6	4	6	5	NT	NT
7123	4	3	7	5	NT	NT
Tolerised						
7039	6	6	5	5	3	3
7040	6	8	3	4	4	3
7042	6	8	10	5	9	7
7074	7	8	8	8	7	5
Test antigen in AACE	AL	FL	AL	FL	AL	FL

[a] Not tested.

of fetal protein. However, this conclusion must be tentative as the method of assaying the immune response was restricted to gel precipitation which may only detect antibodies against multivalent or "strong" antigens. Further, some antigens could have been at a concentration too low for a precipitin arc to form even if a sufficient titre of specific precipitating antibody was present in the antiserum.

4.2 Negative Antibody Affinity Chromatography

Specific proteins can be partially purified by immunoadsorption of contaminating proteins to an antibody affinity column, so that an enriched fraction of specific protein can pass unretarded through the column. This is "negative antibody affinity chromatography" (NAAC), a procedure frequently used to remove albumin from preparations of alpha fetoprotein (e.g. PIHKO et al., 1973) and which has also been used to fractionate soluble tumour antigens (DE CARVALHO et al., 1964), placental extracts (ANDERSON et al., 1974) and amniotic fluid and fetal serum (SUTCLIFFE, 1975; SUTCLIFFE et al., 1978a). We removed adult-type serum protein from amniotic fluid and fetal and maternal serum on 1.9-2.5 kg columns of sepharose 4B (column diameter 15 cm) on which had been immobilised 20-35 g of immunoglobulin from sheep antisera to adult human male serum. Unadsorbed protein (1%-6% of the total applied) was concentrated by ultra-filtration (Amicon PM-10), analysed for protein content and used to immunise New Zealand white rabbits (Table 4.2). Adsorbed protein was removed with 0.5 M acetic acid and the column regenerated with phosphate buffered saline. The columns can be used 20 - 30 times without serious loss of capacity.

Table 4.2. Yields of protein in negative antibody affinity chromatography (NAAC) experiments

Material	Total protein fractionated	Yield of unadsorbed protein
	mg	mg
Amniotic fluid	1480	34
Term umbilical cold serum	4560	57
Term maternal serum fraction	13300	545

[a] Delivery serum (see text) was fractionated in neutral 50% saturated $(NH_4)_2SO_4$ and the precipitate was re-suspended in PBS and used in all NAAC experiments.

4.2.1 Amniotic Fluid

Rabbit anti-sera to the protein fraction of amniotic fluid which was not adsorbed to the NAAC column still showed strong reactions with adult type serum proteins (SUTCLIFFE, 1975). However, when these anti-sera were adsorbed with adult human serum apparently novel antigens were revealed by AACE. Five of these antigens were common to adult and fetal tissues and were at maximum concentration in amniotic fluid during the second trimester of pregnancy. Two other antigens were apparently located only in fetal skin and were not present in adult skin or other tissues. The identity of these antigens is uncertain, though immuno-fluorescence studies by KLAVINS and his colleagues (1971) have also indicated that fetal-specific antigens are detectable in fetal skin.

Rabbit antisera to NAAC-treated amniotic fluid reacted with a further antigen which initially appeared to be specific for the uterine decidua and uterine endometrium. These antibodies could not be adsorbed with extracts of adult liver or adult kidney. The antigen was provisionally termed "alpha-2 uterine protein" (AUP, SUTCLIFFE et al., 1978a). The nature of this protein is discussed below.

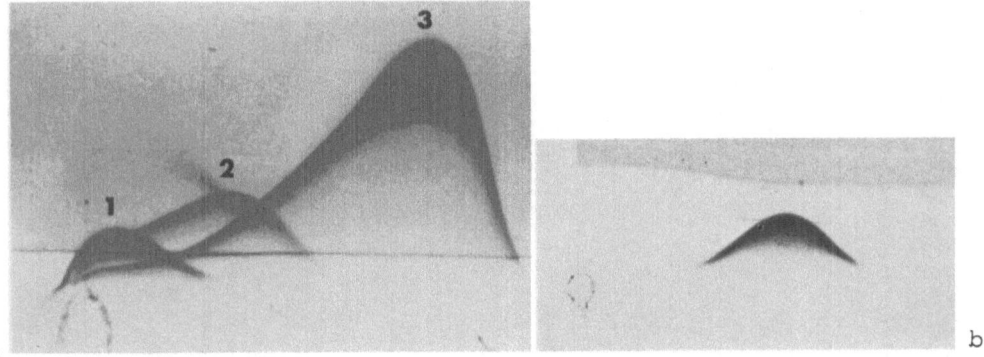

Fig. 4.1 a,b. Antibody-antigen crossed electrophoresis used to test antisera from two rabbits immunised with different preparations of protein from umbilical cord serum, after fractionation by negative affinity chromatography. The antigen wells are seen as oval marks to the left of each plate, the first dimensional anode lies to the right. The antisera had been adsorbed with limiting quantities of adult human serum and are present in the second-dimension at a concentration of 6%. In both plates the antigen tested was 22 µg of the unadsorbed protein from the NAAC experiment (Table 4.2) using term umbilical cord serum as starting material
Peak 1 sex-hormone binding globulin; Peak 2 PAPP-A; Peak 3 AFP

4.2.2 Term Umbilical Cord Serum

The removal of adult-type serum proteins from pooled umbilical cord serum by NAAC provided a fraction which elicited only a very weak antibody response against adult serum proteins when injected into New Zealand white rabbits. However, AACE showed precipitin arcs corresponding to sex-hormone-binding globulin, PAPP-A and α-fetoprotein (Fig. 4.1a). One rabbit produced an antiserum (Fig.4.1b) which reacted with PAPP-A alone (see Sect. 4).

4.2.3 Term Maternal Serum

A pool of maternal serum (1.4 litre) was made from blood collected per vaginam at normal deliveries. This material was contaminated with a small proportion of fetal serum. Serum protein was precipitated with 50% (NH4)$_2$SO4 (to exclude albumin) and the redissolved precipitate was passed through an NAAC column in aliquots of 500-1,700 mg. Unadsorbed material was concentrated, and assayed for total protein, albumin, AFP, hPL, SP-1, PAPP-A, and SP-3 (Table 4.3). Recoveries differed widely between individual proteins but were consistent for different runs on the same NAAC column. The recovery of placental protein was substantially less than that for AFP and may indicate loss of antigen due to instability, to minor antibody specificities on the affinity column or to non-specific interaction with the solid phase. The presence of antibodies to SP-3 on the column, due to low levels of SP-3 in most adult human sera (HORNE et al., 1976), probably explains the low recovery of this protein.

Table 4.3. The recovery of pregnancy protein from negative antibody affinity chromatograpgy (NAAC) [a]

Experiment	Folin protein		Percentage recovery of specific proteins[b]					
	Applied to NAAC mg	Recovered unadsorbed %	Albumin	AFP	hPL	SP-1	PAPP-A	SP-3
1	1232	44 (3.6)	2.6	100	17	24	51	3.4
2	1232	56 (4.5)	2.2	116	42	38	58	7.2
3	1232	46 (3.7)	3.6	89	20	20	54	5.8
4	678	46 (6.8)	6.4	100	31	38	54	6.0

[a] The concentration of individual proteins in the protein applied to the NAAC column were as follows: total protein: 62 mg/ml; albumin: 3.8 mg/ml; AFP: 3.2 µg/ml.

[b] Calculated as: $\dfrac{\text{quantity of protein in unadsorbed fraction}}{\text{quantity of protein applied to NAAC}} \times 100.$

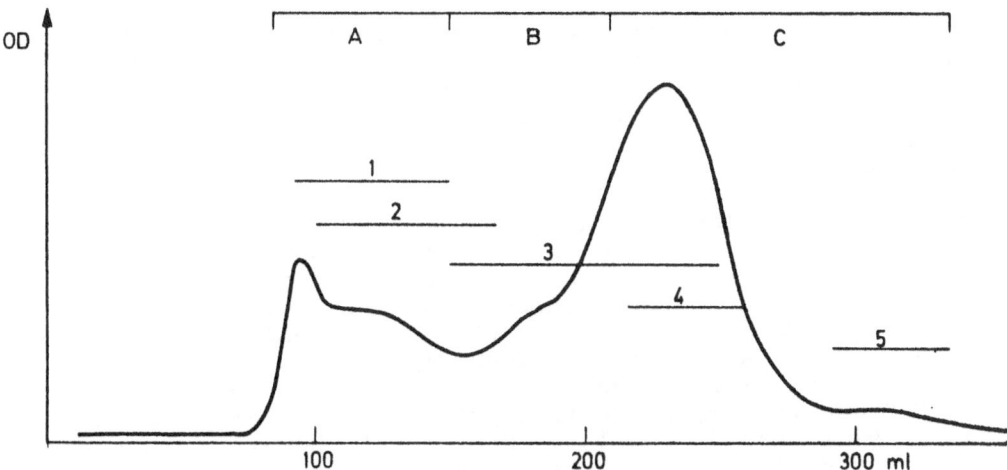

Fig. 4.2. Gel filtration of protein obtained from maternal serum after ammonium sulphate precipitation and fractionation over NAAC columns to remove adult-type serum protein (see Sect. 2.3 and Table 4.2). 532 mg protein in 18 ml phosphate-buffered saline (PBS) was applied to a column (91x2.6 cm) of Ultrogel AcA34 (LKB) at 4° C. The column was washed with PBS at 20 ml/h and fractions were collected at the rate of 3 per h. The optical density was measured using a Uvicord II at 280 nm, and the trace is shown as a heavy fluctuating line. The *short horizontal lines* show the elution ranges of five specific proteins: PAPP-A (1); SP-3 (2); SP-1 (3); AFP (4) and hPL (5). These ranges were determined by one-dimensional antibody-antigen crossed electrophoresis using appropriate mono-specific antisera. The yields of protein in the three large fractions A,B,C (shown in the top of the figure) were: 114 mg (*A*); 162 mg (*B*); 224 mg (*C*). A proportion of the optical density observed between 80 and 100 ml of eluate was due to turbidity

Unadsorbed protein from the NAAC column was further purified by gel filtration on AcA34 (Fig. 4.2). Fractions A, B and C were concentrated and used to immunise New Zealand white rabbits. These yielded antisera to PAPP-A, and AFP respectively. Antibodies to fraction A did not react with any material comparable to PAPP-B.

4.3 The Sensitivity of Pregnancy-Associated Proteins to Dissociants of Antibody-Antigen Complexes

Antibody affinity chromatography is the method of choice for the purification of proteins which are at a low concentration in the initial extract, or associated with a protein contaminant with similar physical properties. For example, though there is a primary sequence homology between albumin and α-fetoprotein (RUOSLAHTI and TERRY, 1976) and they are physico-chemically very similar, they can be readily separated by affinity chromatography using antibodies specific for one or the other component (see SEPPÄLÄ and RUOSLAHTI, 1976).

The efficiency with which antigen is obtained from antibody affinity chromatography is affected by the dissociation constant

of the antibody and by the stability of antigen and antibody
during the purification protocol. A low dissociation constant
effectively results in irreversible binding of antigen to antibody. This can be countered by using low affinity antibodies
taken from animals soon after the onset of the humoral response
(HUNTER, 1976).

Providing the disulphide bonds are not disrupted, antibodies are
relatively stable in a wide variety of strong reagents which dis-
sociate antibody-antigen complexes. Such reagents include the
chaotropic thiocyanate and iodide ions at between 1.5 M and
3.0 M (DANDLIKER et al., 1967, 1968; DE SAUSSURE and DANDLIKER,
1969; and TERNYNCK, 1969); 5 M guanidine-HCl (DANDLIKER et al.,
1968); 8 M urea (SLOBIN and SELA, 1965); Glycine-HCl buffer
(AVRAMEAS and TERNYNCK, 1969); Glycine-NaOH buffer pH11 (OMENN
et al., 1970); and 3 M MgCl$_2$ (HOBBS, 1976). However, many proteins are sensitive to at least some of these dissociating
agents.

We have studied the sensitivity of a variety of pregnancy-associated and other serum proteins by adding different dissociating
agents to 1-2 ml aliquots of pregnancy serum, then removing the
agents by dilution and dialysis before testing for the presence
of immuno-reactive protein by AACE. Low pH, 5 M guanidine-HCl,
8 M guanidine-HCl, 8 M urea and high molarities (2.5 M and 3M)
of thiocyanate and iodide often caused denaturation of antigens,

Table 4.4. Effect of dissociating agents on AACE precipitin arc[a]

Protein	KCNS[b]	KI[b]	5 M guanidine HCL	8 M Urea	pH 2.8
β_1-lipopro- tein	2.0 M	3.0 M	-	-	-
α_2-macro- globulin	1.5 M	2.0 M	-	-	-
PAPP-A	1.5 M	2.0 M	-	-	-
SP-3	-	1.5 M	-	-	-
α_1-lipo- protein	3.0 M	3.0 M	-	-	-
Plasminogen	-	1.5 M	-	-	-
SP-1	-	2.0 M	-	-	-
Albumin	3.0 M	3.0 M	+	+	+
SP-2	1.5 M	2.5 M	-	-	-
AUP	3.0 M	3.0 M	+	+	+
hPL	3.0 M	3.0 M	+	+	+

[a] - = alteration or abolition of precipitin arc;
+ = normal precipitin arc formed.
[b] KCNS and KI each tested, at neutral pH, at 1.5, 2, 2.5 and 3 M in each
case. The molarity noted in the table was the highest molarity which resulted in a normal precipitin arc.

though some proteins (albumin, hPL and AUP) were unaffected.
In some cases denaturation resulted in a complete loss of anti-
genicity as assessed by AACE. In other cases a precipitin arc
could still be detected, but with altered morphology (Table 4.4),
for example, blurring or elongation of the arc, loss of the
cathodal ends at the end of the gel, and the presence of a second arc of variable size and intensity. The type of alteration
depended on the particular antigen rather than on the type of
dissociant. With the exception of certain changes in the precipitin arc of α-2- macroglobulin, none of these minor changes
in morphology could be detected by Ouchterlony double immunodiffusion.

4.4 The Partial Purification of PAPP-A

The finding (Table 4.4) that PAPP-A in maternal serum was sensitive to 3 M KSCN but apparently insensitive to 1.5 - 2 M KI
was confirmed in initial attempts to purify the antigen. A
column of Sepharose coupled with rabbit anti-PAPP-A (Fig. 4.1b)
adsorbed PAPP-A from maternal serum and yielded immunologically
active PAPP-A when treated with 1.5-2 M KI, but not with 3 M
KSCN. Rabbit and sheep antisera resulting from immunisation with
PAPP-A dissociated from solid phase antibodies with 2 M KI showed
specificity identical with an antiserum kindly provided by Lin
and Halbert, and with the antibody shown in Fig. 4.1b.

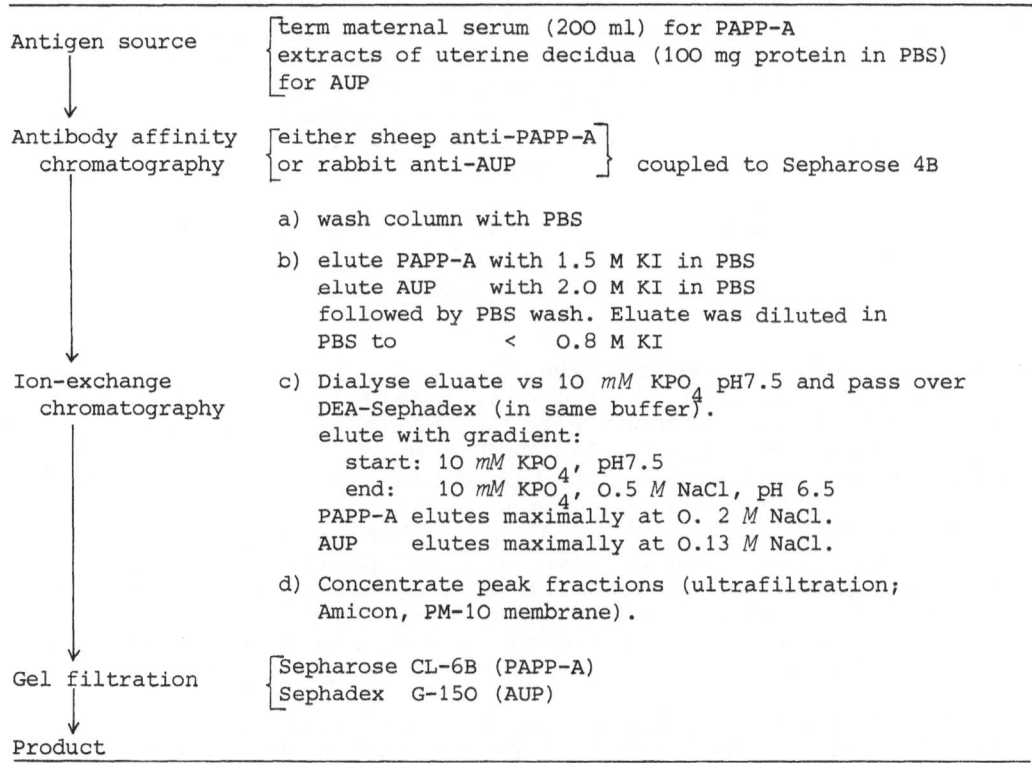

Fig. 4.3. Purification protocols for PAPP-A and AUP

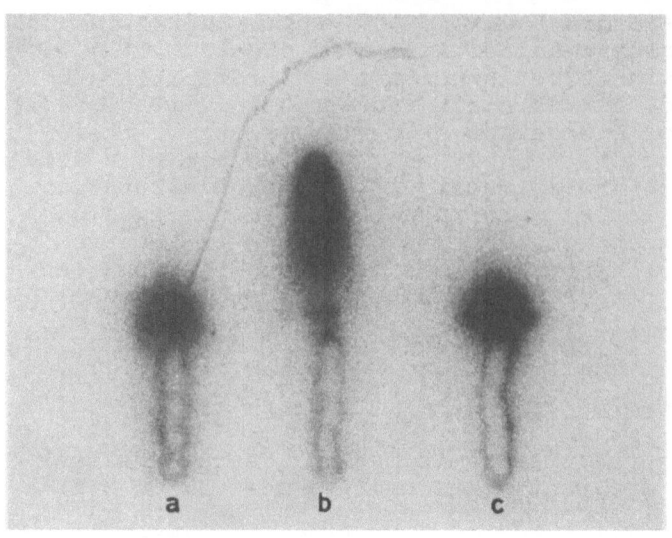

Fig. 4.4. One dimensional antibody-antigen crossed electrophoresis for the analysis of ^{125}I-labelled PAPP-A by autoradiography. The antiserum used was 3% sheep anti-PAPP-A. ^{125}I-labelled PAPP-A (10 µl in 50 mM KPO_4, pH 7.5: 800 cps) was added to each of wells a, b, and c. In addition, well b contained 10 µl term maternal serum and well c contained 10 µl pooled non-pregnant human serum. A similar set of wells were used in a separate (control) plate which contained 3% non-immune serum (NS). After electrophoresis and two days of washing to remove non-precipitated counts, the gels were subjected to autoradiography with Kodirex X-ray film. The plate above shows the result of using a specific sheep antiserum against PAPP-A. The control plate (3% NS) showed only faint activity around the margins of the wells. The areas of intense radioactivity (and the corresponding areas in the control plate) were cut out, counted in a γ-counter and expressed as a percentage of the counts applied.

The results were:

Plate	a	b	c
3% anti-PAPP-A	45 %	49 %	46%
3% NS	2.3%	2.6%	–

A further purification of PAPP-A was attempted using the three-step protocol outlined in Fig. 4.3. Initially, the solid phase sheep antibody was desorbed with 2 M KI in phosphate buffered saline (PBS), but this yielded PAPP-A which appeared to be denatured since it formed blurred precipitin arcs on one-dimensional AACE. The procedure was then repeated with 1.5 M KI in PBS and yielded a preparation of PAPP-A with normal precipitin arc morphology. The difference between this and previous experiments was probably due to the duration of exposure of PAPP-A to 2 M KI. This was longer in the second experiment (Fig. 4.3) in which the column had a volume of 170 ml, and KI was removed from the eluate by dialysis.

PAPP-A purified by the three-step protocol (Fig. 4.3) using 1.5 M KI was iodinated with ^{125}I by the lactoperoxidase technique (BOLTON, 1977). Immunoreactive ^{125}I-PAPP-A was identified as an AACE peak (Fig. 4.4). The precipitin arc contained 45%-49% of

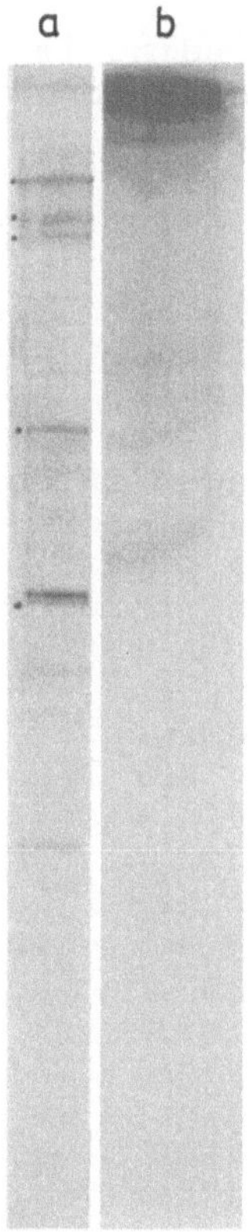

Fig. 4.5. Analysis of immune precipitated ^{125}I PAPP-A by 7%
SDS polyacrylamide slab gel electrophoresis, see section DII
for details. Track (*a*) shows the mobility of a set of five
calibrated viral polypeptides, which are marked by black dots
to the left of the track. Their molecular weights are 156,000,
136,000, 120,000, 66,000 and 42,000 respectively. In track
(*b*) is ^{125}I-labelled PAPP-A which has been dissociated from
anti-PAPP-A coupled to sepharose by boiling in the presence
of SDS and 2 mercaptoethanol. The origin and cathode are at
the top of the gel. Part of the loading gel has been excluded
from the plate. The horizontal line (marked by a dot to the
right of the gel) shows the interface between the loading
and the running gels. We acknowledge the help of Helen Moss
in this experiment

the counts applied to the AACE plate. After purification with
anti-PAPP-A coupled to Sepharose 6B the tracer was analysed by
10% SDS polyacrylamide slab gel electrophoresis (MARSDEN et al.,
1976) (Fig. 4.5). Though the exclusion limit of the gel was
150,000 daltons, PAPP-A tracer remained at the junction of the
loading and the running gels indicating a subunit size of greater than about 150,000 daltons and the possibility that it may
contain more than one covalently bonded subunit.

Since PAPP-A is a glycoprotein, (LIN et al., 1974) an estimate
of carbohydrate content is now required in order to estimate the
polypeptide chain length of the protein.

4.5 Heterogeneity of PAPP-A

AACE has consistently revealed that PAPP-A forms two precipitin
arcs, one within and parallel to the other. This is not detectable by double immunodiffusion. The proteins which form these
arcs are not separable by electrophoresis in two-dimensional
AACE, or by DEAE-sephadex ion exchange or sepharose 4B gel
filtration. This double peak phenomenon has been observed for
partially purified PAPP-A as well as for the protein in maternal serum before any purification. The significance of this

finding is unknown but it suggests that some degree of antigenic heterogeneity exists in the protein.

4.6 Purification of Alpha-2 Uterine Protein (AUP)

AUP has been purified from both uterine decidua and amniotic fluid by antibody affinity chromatography. Compared with PAPP-A, AUP is much more stable in most dissociants of antibody-antigen complexes (Table 4.4) and little difficulty was experienced in obtaining a highly purified preparation of AUP from a sample of uterine decidua from a pregnancy terminated at 11 weeks of gestation (SUTCLIFFE et al., 1978b). The method was similar to the three-step procedure shown in Fig. 4.3 except that the protein was eluted in 2 M KI and that gel filtration was carried out on G-150 sephadex. After iodination 60% of the radiolabelled AUP was precipitated in a double antibody system and in AACE. On SDS electrophoresis the pattern of the radio-iodinated protein was similar to that of the unlabelled material. Both showed a major component of between 23,000-25,000 daltons. Since G-150 sephadex gel filtration consistently showed that AUP from decidua or amniotic fluid had a molecular weight of 50,000 daltons, it was concluded that the protein is a dimer composed of two very similar or identical polypeptide chains (SUTCLIFFE et al., 1978b). The purified material was readily iodinated and could be used as the basis of a double antibody radioimmunoassay which gave results similar to those obtained by one-dimensional AACE.

4.7 Cross-Reactivity and Tissue Distribution of AUP

Using Ouchterlony double immunodiffusion and AACE, we have been unable to demonstrate cross-reactivity between AUP and SP-1, SP-3, hPL, hCG, AFP, CEA, Ferritin, PAPP-A and adult male serum proteins. AUP is not detectable by AACE in maternal serum at any stage of gestation. It is detectable in amniotic fluid by AACE where it is at maximum concentrations between 14 and 18 weeks of pregnancy, although we have no knowledge of the concentration of AUP in amniotic fluid before 12 weeks of pregnancy because of unavailability of material (SUTCLIFFE et al., 1978a).

Antisera have been raised against AUP obtained from amniotic fluid and decidua, and these cross-react completely in double immunodiffusion (Fig. 4.6). The molecular weights of AUP from the two sources are indistinguishable (50,000 daltons). AUP can be detected in the endometrium of the non-pregnant uterus but not in adult liver, gut, kidney, lung or skin. AUP has been detected and quantitated in decidual tissue from between 8 and 13 weeks of pregnancy (Table 4.5) and at delivery (Table 4.6). The concentrations of AUP in decidua early in pregnancy were found to be between 50 and 100 times greater than those in decidua at between 35 and 40 weeks. Although at term no AUP is detectable in the placenta, we have recently found considerable quantities in placentae between 8 and 13 weeks (Table 4.5).

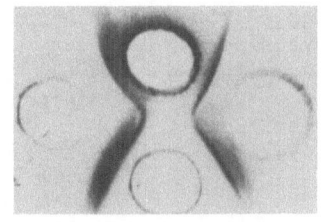

Fig. 4.6. Ouchterlony immunodiffusion of antisera raised against AUP of different origins. Contents of wells, starting at the top and going clockwise: rabbit anti-AUP from amniotic fluid (SUTCLIFFE et al., 1978a); 20 µl homogenate of 11-week decidua at protein concentration of 4 mg/ml; rabbit anti-AUP from uterine decidua of similar gestational age; finally (left-hand well), 20 µl 16-week gestation amniotic fluid

The consistent presence of AUP in uterine tissue suggests that AUP is synthesised in uterine tissue, and that its measurement might be of value in the diagnosis of pathological conditions of the uterus. However, the significance of the AUP detected in early placenta is difficult to assess. Since the placenta is closely adherent to the uterus it is possible that the AUP detected in early placental material was in fact present in associated cells of uterine origin. This question may be resolved by immunohistochemistry and the availability of purified AUP will allow appropriate adsorption controls. The relatively high concentrations of AUP in decidua in early pregnancy may account for the high levels of AUP found in amniotic fluid between 14 and 18 weeks. However, while an approximately ten-fold decline in fluid levels of AUP has been observed as gestation proceeds (SUTCLIFFE et., 1978a), the decline in the AUP content of decidua is somewhat greater, in the range of 1/50 to 1/100 of that observed early in pregnancy (Tables 4.5 and 4.6). It is also interesting that the maximal concentration of AUP in amniotic fluid is coincident with the maxima for the specific activities of α-1,4 glucosidase and heat-stable alkaline phosphatase (see SUTCLIFFE, 1976). Thus, all three proteins may enter the amniotic fluid from either the uterus or the placenta, though there are some difficulties in visualising a placental contribution. If the placental villi are completely bathed in maternal blood then proteins from the tissue would enter the maternal blood and reach the amniotic fluid indirectly in a manner similar to that of maternal serum proteins of comparable size. This clearly does not occur for AUP (50,000 MW) since it is not detectable in maternal serum by AACE and its concentration in amniotic fluid falls well before that of serum albumin (68,000 MW) or of group-specific component (45,000 MW, and known to be of maternal serum origin, see SUTCLIFFE, 1976). However, if a proportion of the placenta is not freely perfused by maternal blood but instead lies in close apposition to the chorionic plate, then it is possible that there is a placental contribution to the AUP present in early amniotic fluid.

Table 4.5. Concentration of AUP in decidual tissue and placenta early in pregnancy

Gestation (weeks)	Concentration of AUP	(Units/mg protein)[a]
	Decidual tissue[b]	Placenta[b]
8	NT[d]	0.2
8	2.8	NT
9	0.9	0.2
11	2.2	2.0
13[c]	1.4, 0.4	NT
13	0.6	0.5

[a] AUP assayed by one dimensional AACE (SUTCLIFFE et al., 1978a).

[b] Material obtained from vacuum termination of pregnancy by kind permission of Dr. W.R. Chatfield and Dr. F. Sharp.

[c] Two separate samples of decidua were recovered from this case and two assayed separately as shown.

[d] Not tested.

Table 4.6. Concentration of AUP in decidual tissue at delivery

Gestational period (weeks)	Number of patients	AUP concentration	(Units/mg protein)[a]
		Range	Mean
35–37	3	0.010 –0.020	0.015
38–39	7	0.0025–0.085	0.027
40	4	0.004 –0.010	0.0065

[a] See footnotes a and b to Table 4.5. Delivery was carried out by elective caesarian section.

References

Anderson, N.G., Holladay, D.W., Caton,J.E., Candler, E.L., Dierlam, P.J., Eveleigh, J.W., Ball, F.L., Hollemann, J.W., Breillatt, J.P., Coggin, J.H.: Searching for human tumour antigens. Cancer Res. 34, 2066-2076 (1974)

Avrameas, S., Ternynck, T.: The cross-linking of protein with glutaraldehyde and its use for the preparation of immunoadsorbents. Immunochemistry 6, 53-66 (1969)

Bohn, H.: Isolation and characterisation of placental specific proteins SP-1 and PP-5. Protides Biol. Fluids 24, 117-124 (1976)

Bolton, A.E.: Radioiodination techniques. In: Reviews of the radiochemical centre, Vol. XVII, pp. 45-58. Amersham: Radiochemical Centre 1977

Dandliker, W.B., Alonso, R., de Saussure, V.A., Kierszenbaum, F., Levison, S.A., Shapiro, H.C.: The effect of chaotropic ions on the dissociation of antigen-antibody complexes. Biochemistry 6, 1460-1467 (1967)

Dandliker, W.B., de Saussure, V.A., Levandoski, N.: Antibody purification at neutral pH using immunospecific adsorbents. Immunochemistry 5, 357-365 (1968)

De Carvalho, S., Lewis, A.J., Rand, H.J., Uhrick, J.R.: Immunochromatographic partition of soluble antigens. Nature 204, 265-266 (1964)

De Saussure, V.A., Dandlicker, W.B.: Ultracentrifuge studies of the effects of thiocyanate ion on antigen-antibody systems. Immunochemistry 6, 77-83 (1969)

Gold, P., Freedman, S.O.: Demonstration of tumour-specific antigens in human colonic carcinomata. J. Exp. Med. 121, 439-462 (1965)

Hobbs, J.R.: Affinity purified antibodies enabling immunoperoxidase reactions. Communicated at the 24th meeting of protides of the biological fluids 1976

Horne, C.H.W., Bohn, H., Towler, C.M.: Pregnancy-associated α_2-glycoprotein. In: Plasma hormone assays in the evaluation of fetal wellbeing. KLOPPER, A. (ed.). Edinburgh: Churchill Livingstone 1976

Hunter, W.M.: Radioimmunoassay. In: Immunochemistry. WEIR, J. (ed.), Vol. I, pp. 14.1-14.40. Oxford: Blackwell 1976

Klavins, J.V., Mesa-Tejada, R., Weiss, M.: Human carcinoma antigens cross reacting with anti-embryonic antibodies. Nature New Biol. 234, 153-154 (1971)

Lin, T.M., Halbert, S.P., Kiefer, D., Spellacy, W.N: Three pregnancy associated human plasma proteins. Int. Arch. Allergy Appl. Immunol. 47, 35-53 (1974)

Marsden, H.S., Crombie, I.K., Sharpe, I.H.: Control of protein synthesis in Herpes virus-infected cells: analysis of the polypeptides induced by wild type and sixteen temperature-sensitive mutants of HSV strain 17. J. Gen.Virol. 31, 347-372 (1976)

Omenn, G.S., Ontjes, D.A., Anfinsen, C.B.: Fractionation of antibodies against Staphylococcal Nuclease on sepharose immuno-adsorbants. Nature 225, 189-190 (1970)

Pihko, H., Lindgren, J., Ruoslahti, E.: Rabbit α-fetoprotein: immunochemical purification and partial characterisation. Immunochemistry 10, 381-385 (1973)

Ruoslahti, E., Terry, W.D.: Alpha foetoprotein and serum albumin show sequence homology. Nature 260, 804-805 (1976)

Seppälä, M., Ruoslahti, E.: Alpha-fetoprotein. In: Contributions to gynecology and obstetrics. KELLER, P.J. (ed.), pp. 144-186. Basel: Karger 1976

Slobin, L.I., Sela, M.: Use of urea in the purification of antibodies. Biochem. Biophys. Acta 107, 593-596 (1965)

Sutcliffe, R.G.: The nature and origin of the soluble protein in human amniotic fluid. Biol. Rev. 50, 1-33 (1975)

Sutcliffe, R.G.: The search for new human fetal proteins. Protides Biol. Fluids 24, 543-546 (1976)

Sutcliffe, R.G., Brock, D.J.H., Nicholson, L.V.B., Dunn, E.: Fetal and
   uterine-specific antigens in human amniotic fluid. J. Reprod. Fertil. <u>54</u>
   (1978a) (in press)

Sutcliffe, R.G., Bolton, A.E., Sharp, F., Nicholson, L.V.B., Mackinnon, R.:
   The purification of human alpha uterine protein (AUP) by antibody affinity
   chromatography (Submitted) (1978b)

5 Isolation and Characterization of Placental Proteins with Special Reference to Pregnancy-Specific β_1 Glycoprotein and Other Proteins Specific to the Placenta

H.Bohn

5.1 Detection and Characterization of Proteins from the Human Placenta

During the past 10 years in our laboratory a number of proteins have been isolated and characterized from human term placentae and this work is being continued. Some of the proteins already isolated can be used therapeutically; others have been found to be valuable in pregnancy and/or tumour diagnosis. The present paper reviews this work. The first part describes all the placental proteins so far detected or isolated; the second part deals with the isolation, characterization and clinical significance of those proteins which are specific to the placenta.

5.1.1 Placental Fibrin-Stabilizing Factor

The first protein purified in our laboratory from placental extracts was a fibrin-stabilizing factor (FSF) (BOHN and SCHWICK, 1971). FSFs play a role in blood coagulation and also in wound healing. The function in coagulation is enzymatic covalent cross-linking of fibrin molecules through lysine and glutamine residues, a process known as fibrin stabilization. The human FSFs are pro-enzymes which are activated by thrombin and converted to active transglutaminases.

Two different FSFs occur in blood; one in plasma, the other in platelets. Fibrin-stabilizing activities have also been detected in human tissues such as aorta and skin. In 1967 we found that extracts from human term placentae also exhibit high fibrin-stabilizing activity. Purification and characterization of FSF from placental extracts showed it to be identical to platelet FSF and closely related to FSF from plasma (BOHN et al., 1972). Table 5.1 summarizes the physicochemical properties and the subunit structure of the three human FSFs.

The molecules of placental and platelet FSF are composed of two identical polypeptide chains (subunit A) each of molecular weight 80,000 daltons, linked by non-covalent bonds. These FSFs are, therefore, described as A_2. The molecules of plasma FSF (factor XIII) also contain two A subunits and in addition another component (subunit S, molecular weight 180,000 daltons), also linked by non-covalent bonds. Plasma FSF can be described as A_2S. Biological activity in all FSFs resides in the common subunit A; subunit S has no fibrin-stabilizing activity. A concentrate of placental FSF (Behringwerke) is now used therapeutically for treatment of patients with factor XIII deficiency and for the acceleration of wound healing after major surgery.

Only a few of the large number of biologically active proteins of the placenta have been purified and compared physicochemically with active substances of other human tissues or body fluids. Yet presumably most of these placental proteins are, like placental FSF, identical or similar to the corresponding proteins in other tissues.

Table 5.1. Physicochemical properties and subunit structure of the human fibrin-stabilizing factors

	Electro- phoretic mobility	Carbo- hydrate content (%)	Molecular weight	Subunit structure
Placental FSF	β_2	1.5	160,000	A_2
Platelet FSF	β_2	1.5	160,000	A_2
Plasma FSF	β_2	4.9	340,000	A_2S

Table 5.2. Placenta-specific hormones and enzymes

Hormones	
hPL	Human placental lactogen
hCG	Human chorionic gonadotrophin
hCT	Human chorionic thyrotrophin

Enzymes	
HSAP	Heat stable alkaline phosphatase
CAP	Cystine amino peptidase (oxytocinase)
17β-HSD	17β-hydroxysteroid dehydrogenase
DAO	Diamine oxidase (histaminase)

5.1.2 Pregnancy Proteins

Only a few of the numerous biologically active compounds found in the placenta appear to be specific to the placenta. These proteins are listed in Table 5.2; some have hormonal functions, other are enzymes.

The placenta-specific hormones and enzymes are synthesized in the trophoblast and secreted into the maternal blood. They are also termed pregnancy-specific proteins because they normally appear in the circulation only during pregnancy. One such protein, hPL, was found in a side fraction from the placental FSF purification. Using this material we obtained hPL, which could be crystallized as a euglobulin after solution in dilute neutral buffer and slowly adjusting the pH to 5.5 (BOHN, 1971a).

The purified hPL was used to prepare specific antisera and to develop methods for the immunochemical determination of this protein. Levels of hPL in maternal serum rise progressively during pregnancy and the measurement of hPL is of diagnostic value in detecting and monitoring high-risk pregnancies. hPL may also have therapeutic applications, since it has lactogenic and growth promoting activities. Administration of hPL to five pregnant women in order to prevent threatened abortion has been reported recently (NERI et al., 1976). In a systematic search for other pregnancy proteins in the placental extracts we immunized rabbits with protein fractions from human placentae and absorbed the resulting sera with normal human serum. Antisera prepared in this way detected four different proteins in sera from pregnant women on Ouchterlony gel diffusion (BOHN, 1971b). One of these proteins was identified as human placental lactogen (hPL). The other three appeared to be different from all known pregnancy-specific hormones and enzymes and were designated as SP_1, SP_2 and SP_3, (SP stands for Schwangerschaftsprotein' or pregnancy protein').

Table 5.3. Physicochemical properties of the pregnancy proteins SP_1, SP_2 and SP_3

	Electro-phoretic mobility	Molecular weight	Carbo-hydrate content (%)
SP_1 Pregnancy-specific β_1 glycoprotein	β_1	90,000	29.3
SP_2 Steroid-binding β globulin	β_1	65,000	12.6
SP_3 Pregnancy-associated α_2 glycoprotein, pregnancy zone protein, serum factor Xh	α_1	360,000	12.2

SP_1 proved to be specific to pregnancy, being found only in sera from pregnant women. SP_2 and SP_3 were not specific, occurring in trace amounts (usually less than 1 mg/100 ml) in all normal sera, but with a striking elevation during pregnancy. These proteins have, therefore been termed 'pregnancy-associated proteins'. The pregnancy proteins SP_1 and SP_3 were isolated from placental extracts; their physicochemical properties are summarized in Table 5.3. All three are glycoproteins, the highest carbohydrate content being found in SP_1.

SP_1 has a carbohydrate content of around 30%, a molecular weight of 90,000 daltons, and the electrophoretic mobility of a β_1 globulin. The name 'pregnancy-specific β_1 glycoprotein was chosen. The biological function of SP_1 is unknown.

SP_2 was identified as the steroid-binding β globulin (sex-hormone-binding globulin, SHBG). This protein binds steroid hormones, especially testosterone and oestradiol, with high affinity.

SP_3 was found to be identical with serum factor Xh, the so-called pregnancy zone protein. This protein is now generally designated as pregnancy-associated α_2-glycoprotein (α_2PA-glycoprotein). The biological function of SP_3 is uncertain, though it apparently has immunosuppressive properties; SP_3 may play a role in preventing rejection of the fetal allograft during pregnancy.

In a collaborative study (TOWLER et al., 1976) we have measured the plasma concentrations of SP_1, SP_2 and SP_3 throughout pregnancy in 15 primigravidae. Levels of SP_1 were found to rise progressively to reach a plateau in the last 4 weeks. SP_2 shows a substantial increase during the first half of pregnancy, and only a slight further elevation in the second half. SP_3 levels reach a maximum by week 35 and thereafter appear to fall slowly until parturition. The highest concentrations found in sera from 15 normal pregnant women (all primigravidae) were as follows: SP_1 34.6 mg/100 ml, SP_2 12.4 mg/100 ml and SP_3 140 mg/100 ml.

After delivery the pregnancy-specific protein SP_1 disappears from the maternal circulation with a half-life of 30-40 h whereas the pregnancy-associated proteins SP_2 and SP_3 return to normal values with a half-life of approximately 6-7 days (BOHN, 1974b).

The increased production of SP_2 and SP_3 during pregnancy is probably induced by steroid hormones, since a similar elevation is seen in women taking hormonal contraceptives. SP_1 is a pregnancy-specific protein and synthesized in the trophoblast. SP_2 and SP_3 are not of placental origin; and the material present in the placenta can be accounted for by the content of maternal blood.

5.1.3 Soluble Placental Tissue Proteins

Rabbits immunized with placental extracts produced antibodies to antigens in these extracts which could not be detected in normal sera from pregnant women. These proteins were soluble constituents of placental tissue which were not secreted into the maternal bloodstream in concentrations sufficient for detection by gel diffusion. Seven such proteins have been characterized in our laboratory, and designated as PP_1, PP_2, PP_3, PP_4, PP_5, PP_6, and PP_7 (BOHN 1973, 1975b; BOHN and KRAUS, 1976; BOHN and WINCKLER, 1977a,b).

The physicochemical properties of these proteins and their concentration in placental extracts are summarized in Table 5.4. The localization of these proteins in the human term placenta by immunofluorescence has been reported by SEDLACEK et al. (1976).

Table 5.4. Characterization of the soluble placental tissue proteins

| Placental protein | Name or function | Physicochemical properties | | | Amounts found in extracts of term placentae |
		Electrophoretic mobility	Molecular weight	Carbohydrate content (%)	mg/placenta
PP_1		α_1	160,000	2.7	3
PP_2	Ferritin	α_2	500,000	-	18
PP_3		α_2	100,000	?	?
PP_4	Albumin	Albumin	60,000	?	?
PP_5	Protease inhibitor	β_1	36,600	19.8	1.5
PP_6		α_1	1,000,000	6.6	100
PP_7		$\alpha_2-\beta_1$	40,000	5.4	60

Table 5.5. Occurrence of the placental tissue proteins in extracts of other human tissues

	PP_1	PP_2	PP_3	PP_4	PP_5	PP_6	PP_7
Fetal tissues							
Heart	-	-	-	-	-	-	++
Kidney	+	-	-	-	-	-	++
Liver	+	-	-	-	-	-	++
Lung	+	+	-	-	-	(+)	++
Stomach	+	-	-	-	-	-	++
Brain	-	+	-	-	-	+	++
Adult tissues							
Heart	-	-	-	-	-	+	++
Lung	-	(+)	-	-	-	-	++
Skin	-	-	-	+	-	-	+
Stomach	-	+	(+)	+	-	+	++
Kidney	+	+	-	+	-	(+)	++
Uterus	-	(+)	-	-	-	-	++
Liver	+	+	-	-	-	+	++
Spleen	(+)	+	-	-	-	+	+
Adrenal	-	-	-	(+)?	-	(+)	++
Colon	-	(+)?	-	-	-	-	++
Rectum	-	(+)?	-	-	-	-	++
Bladder	-	(+)	-	-	-	-	++
Erythrocytes	-	-	-	-	-	(+)	(+)

The soluble placental tissue proteins have the mobility of albumin, α_1-, α_2- or β_1 globulins, and molecular weights ranging from 30,000 to 1,000,000 daltons. The carbohydrate content is less than 10%, with the exception of PP_5 (19.8%). PP_2 has been identified as ferritin, an iron storage protein. PP_5 was found to inhibit plasmin and trypsin, and its biological role may be that of a protease inhibitor. The function of the other placental proteins is unknown. With the exception of PP_5, all placental antigens could be detected in extracts from other human fetal and adult tissues (Table 5.5). PP_7 was present in relatively high concentrations in every tissue examined, and PP_6 and PP_7 were detected in red cell lysates.

Antisera with antibodies to a further 15 soluble placental tissue proteins have now been obtained. Several of these have been purified; some appear to be placenta specific, like PP_5; others occur in a range of normal tissues.

5.1.4 Solubilized Placental Tissue Proteins

The studies reported above concern antigens which can be extracted from the placenta with physiological saline. Equally important materials might exist which are insoluble, i.e. membrane associated.

To investigate these, placental tissue was minced and freed from soluble proteins by extensive washing in physiological saline. Membrane-bound antigens were then dissociated using an acidic buffer with and without addition of mercaptoethanol, chaotropic salt solutions, detergents and digestion with papain (Table 5.6). Solubilized material was separated by centrifugation, neutralized, dialyzed against water or saline, and finally concentrated by ultrafiltration. The final solution was used to immunize rabbits, and the antisera were absorbed with normal human serum.

Using the Ouchterlony double-diffusion technique a number of known proteins were detected, including HSAP, SP_1 hPL, PP_2 (ferritin), PP_4, and certain plasma proteins such as the fibrinogen split products D and E and cold-insoluble globulin; the latter could have been derived from the fibrinoid layer on the trophoblast. In addition, antibodies were detected to at least 11 new, solubilized antigens. These have been designated alphabetically with the letters A-L (Table 5.6). Whether they are specific to the placenta remains to be investigated.

5.2   Proteins Specific to the Placenta

5.2.1   Pregnancy-Specific β_1 Glycoprotein

5.2.1.1 Occurrence

Pregnancy-specific β_1 glycoprotein (SP_1) was first detected by immunochemical methods in sera from pregnant women and in extracts from human placentae. This protein is not normally present in sera from nonpregnant individuals (BOHN, 1971b), nor can it be detected in other normal fetal or adult human tissues (BOHN, 1972b). SP_1 appears to be specific to the placenta. Immuno-histochemical studies have shown that SP_1 is mainly localized in the cytoplasm of the syncytiotrophoblast (BOHN and SEDLACEK, 1975; HORNE et al., 1976).

The mean content of SP_1 in a human term placenta is 30 mg. During pregnancy SP_1 is secreted into the maternal circulation, reaching levels around 15 mg/100 ml at term with a maximum of 35 mg/100 ml (BOHN, 1972a). These are the highest levels of any placenta-specific protein in maternal serum (hPL has a maximum of 1 mg/100 ml and HSAP 0.1 mg/100 ml).

The most abundant specific antigen in placental tissue seems to be hPL, followed by HSAP and SP_1 (Table 5.7). Other pregnancy proteins are present in much lower concentrations (BOHN, 1975a).

In cord blood sera SP_1 is present only in trace amounts. Trace amounts of SP_1 have also been found in colostrum and amniotic

Table 5.6. Detection of solubilized placental tissue proteins

Solubilization procedure	Known antigens to which antibodies were formed	New, solubilized antigens to which antibodies were formed
5% Glycine-HCL pH 2.5	FSP-E, CIG	A, B, E
5% Glycine-HCL pH 2,5 +2% Mercaptoethanol	FSP-E, CIG	B, C
3 M KCL	HSAP	D
2.8 M KI	HSAP, SP_1, PP_4	E, F
6 M Urea	SP_1	E, F, G, I
0,5% SDS	HSAP, hPL	F, K, H, I, L
5% Triton X-100	HSAP, SP_1, PP_2	K, L
Papain	FSP-D, FSP-E	K

KCl, potassium chloride; KI, potassium iodide; SDS, Sodium Dodecyl sulphate; FSP, fibrinogen split product; CIG, cold-insoluble globulin.

Table 5.7. Amounts of placenta-specific proteins in term placentae and in pregnancy sera

| | Term placentae | Pregnancy sera |
	mg per placenta mean values	mg/ 100 ml maximal values
hPL	150	1.0
HSAP	40	< 0.1
SP_1	30	35

fluid (BOHN, 1974a; TATRA et al., 1976). Pregnancy-specific β_1-glycoprotein can also be detected in sera from patients with trophoblastic tumours (hydatidiform mole and chorion-epithelioma), and in the tissue of such tumours (TATARINOV et al., 1974, 1976). The ectopic or inappropriate production of SP_1 by nontrophoblastic malignant tumours has been reported recently (HORNE et al., 1976a, GRUDZINSKAS et al., 1977; TATARINOV and SOKOLOV, 1977; HEYDERMAN 1978; WÜRZ, 1978). Thus, SP_1 can be described as a carcino-placental antigen. A protein showing partial immunochemical identity to serum and placental SP_1 is found in the urine of pregnant women. It has a lower molecular weight and is probably a degradation product of native SP_1 (BOHN, 1972a).

Proteins immunochemically related to SP_1 are present in placentae and pregnancy sera from nonhuman primates but not in animals of lower phylogenetic orders. The corresponding proteins in apes and monkeys are only partly the same as human SP_1 as indicated by spur formation in precipitin arcs. These proteins are also trophoblast-specific and occur only during pregnancy (BOHN and RONNEBERGER, 1973; BOHN 1974a, 1975; STEVENS et al., 1976).

5.2.1.2 Isolation

Pregnancy-specific β_1 glycoprotein (SP_1) was first isolated from an extract of full-term placentae (BOHN, 1971b). The placentae were minced and extracted with 0.4% sodium chloride solution. The extract was then fractionated with 2-ethoxy-6.9-diamino-acridine lactate (Rivanol, Hoechst AG) and ammonium sulphate to give six protein fractions (placental fractions I-VI). SP_1, which was mainly concentrated in placental fraction V, was further purified by gel filtration, ion exchange chromatography on DEAE-cellulose and DEAE-Sephadex, and preparative zone electrophoresis (Table 5.8). The preparation thus obtained was more than 95% pure and the yield was about 5% (BOHN, 1972a). Later it was found that SP_1 could be obtained in good yield and in a highly purified form by use of an immunoadsorption technique in combination with chromatography on hydroxyapatite (BOHN et al., 1976). In this case the placental extract was fractionated with Rivanol and ammonium sulphate to give three protein fractions (placental fractions 1-3). Placental fraction 2, which contained the SP_1, served as starting material for the immuno-

Table 5.8. Isolation procedures for SP_1

A	B
Placental fraction V	Placental fraction 2
DEAE-cellulose	Immunoadsorption
Sephadex G-150	Hydroxyapatite
Zone electrophoresis	
DEAE-sephadex	
Sephadex G-150	
Yield 5%	Yield 50%
Purity > 95%	Purity > 99%

adsorption step. The preparation then obtained was purified by
chromatography on hydroxyapatite; SP_1 dissolved in 0,005 M phosphate buffer passes through hydroxyapatite unretarded, whereas
impurities present in the crude preparation are retained on the
column. With this procedure the yield was 50% and the purity
more than 99% (Table 5.8).

The immunoadsorption technique has also been used to isolate
the proteins antigenically related to SP_1 from placental extracts of apes and monkeys and from urine of pregnant women
(BOHN et al., 1976; BOHN and KRAUS, 1977).

The electrophoretic mobility of human SP_1 on polyacrylamide gel
is slower and the protein is less acidic than the corresponding
proteins from chimpanzee, rhesus and cynomolgus monkeys
(Fig. 5.1). The electrophoretic migration of SP_1 and the corresponding protein isolated from pregnancy urine in polyacrylamide gel containing sodium dodecyl sulphate yields estimated
molecular weights for SP_1 of 90,000 and for urine-SP_1 of 65,000
daltons (Fig. 5.2).

5.2.1.3 Characterization

Table 5.9. summarizes the physical properties of human SP_1 from
placental extracts and urine and the comparable material from
rhesus monkey placentae. The molecular weights were determined
in polyacrylamide gel containing sodium dodecyl sulphate, using
human albumin and its aggregates as markers. Rhesus-SP_1 had two
components of different molecular sizes (80,000 and 110,000 daltons). Reduction with mercaptoethanol did not reduce the molecular size of human SP_1, rhesus-SP_1 or urine-SP_1; this indicates
that the molecules are composed of a single peptide chain. All
three proteins are rich in carbohydrate; the highest value was
found in rhesus-SP_1 (Table 5.10).

Aspartic acid, leucine, serine and glutamic acid are the most
abundant amino acids in human SP_1, and the amino acid composi-
tions of rhesus-SP_1 and urine-SP_1 are similar (Table 5.11).

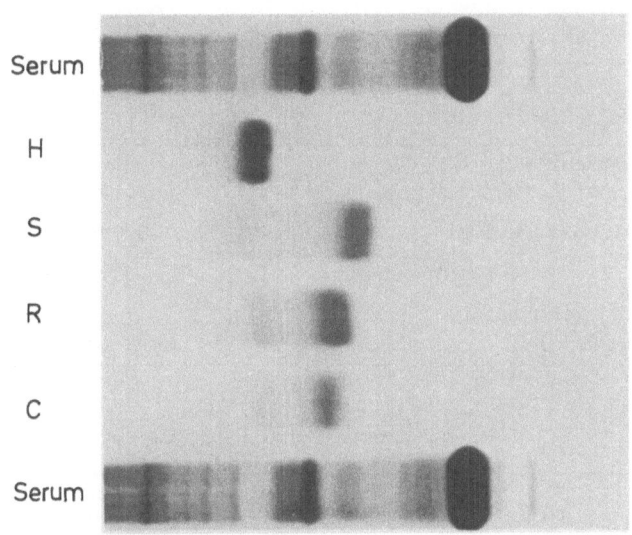

Fig. 5.1. Electrophoretic separation of SP_1 preparations from man and non-human primates in normal polyacrylamide gel. H, Human; S, chimpanzee; R, rhesus; C, cynomolgus

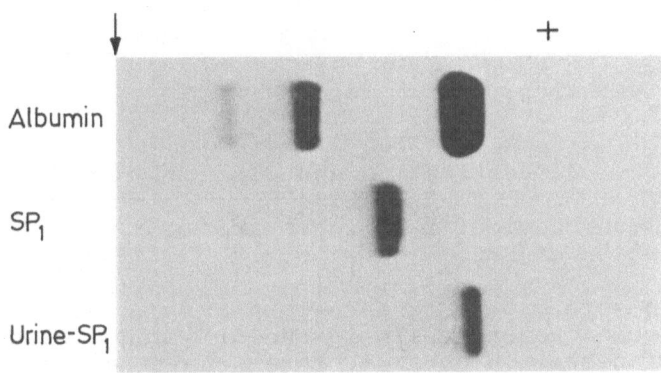

Fig. 5.2. Electrophoretic separation of human SP_1 and urine SP_1 in polyacrylamide gel containing sodium dodecylsulphate. Albumin and its aggregate served as markers

In human SP_1 histidine is the N-terminal amino acid (BOHN et al., 1976). No sequence studies have as yet been performed.

The molecules of human SP_1 were found to be heterogeneous in charge (BOHN, 1973a). Crossed immunoelectrophoresis also revealed differences in antigenic determinants which could be due to different genetic variants of SP_1. The corresponding proteins from apes and monkeys show even greater heterogeneity than human SP_1 (BOHN and SEDLACEK, 1975; BOHN et al., 1976).

Table 5.9. Physical properties

	Human SP$_1$	Rhesus SP$_1$	Human urine SP$_1$
Sedimentation coefficient ($s_{20,w}$)	4.5 S	3.8 S	2.9 S
Molecular weight	90,000	80,000	65,000
Isoelectric point	4.1	3.8	3.8
Electrophoretic mobility	β_1	β_1-α_2	β_1

Table 5.10. Carbohydrate content of placental proteins (mg/100 mg)

	Human SP$_1$	Rhesus SP$_1$	Human urine-SP$_1$
Hexoses	11.7	13.0	11.6
Hexosamines[a]	10.2	12.6	6.4
Fucose	0.6	1.1	0.7
Neuraminic acid[a]	6.8	7.8	6.7
	29.3	34.5	25.4

[a] As N-acetyl derivatives.

5.2.1.4 Quantitation and Clinical Significance of SP$_1$

Immunochemical techniques are used for the quantitative determination of SP$_1$.

In sera from pregnant women SP$_1$ can be measured with the Laurell technique or by single radial immunodiffusion from the 10th week of gestation onwards. More sensitive methods are necessary to detect and quantitate SP$_1$ in early pregnancy sera, in cord blood sera, in amniotic fluid, in the urine of pregnant women and in sera from patients with tumours. To this end radioimmunoassays (GRUDZINSKAS et al., 1977a) and enzymoimmunoassays (GRENNER, 1978) for SP$_1$ have been developed.

The first clinical studies were reported by BOHN (1972a) using radial immunodiffusion. TATRA et al., (1974), using partigen plates (Behringwerke AG) determined the normal range for maternal plasma SP$_1$ levels in a large number of subjects. They demonstrated (TATRA et al., 1974, 1976) that the levels of SP$_1$ in maternal serum correlated directly with placental and fetal weight and fell outside the normal range in abnormal pregnancies. Similar findings have been reported recently by others (GORDON et al., 1977; TOWLER et al., 1977). Measurement of SP$_1$ now appears to provide a new index of fetal well-being.

Table 5.11. Amino acid content of placental proteins (residues per 100 residues)

	Human SP$_1$	Rhesus SP$_1$	Human urine-SP$_1$
Lysine	4.37	5.45	3.21
Histidine	1.77	2.06	2.00
Arginine	5.07	3.61	4.09
Aspartic acid	9.85	9.48	8.71
Threonine	8.24	8.72	8.40
Serine	9.07	10.27	9.30
Glutamic acid	8.97	8.70	9.47
Proline	7.96	6.52	7.19
Glycine	7.07	6.94	9.83
Alanine	3.74	4.02	5.18
Cystine/2	1.57	1.55	1.83
Valine	5.84	5.76	4.88
Methionine	0.88	0.69	1.17
Isoleucine	5.79	6.47	6.04
Leucine	9.60	9.03	9.63
Tyrosine	5.75	6.37	5.23
Phenylalanine	2.12	2.59	1.60
Tryptophan	2.39	1.80	2.19

Using a sensitive radioimmunoassay SP$_1$ can be detected in peripheral blood shortly after conception, and may become valuable as a new "pregnancy test" (GRUDZINSKAS et al., 1977a,b). The occurrence of SP$_1$ in sera from patients with trophoblastic tumours suggests its use as a tumour marker. TATARINOV and SOKOLOV (1977), using a radioimmunoassay, detected SP$_1$ in 76.7% of cases with trophoblastic tumours and in 15% of cases with non-trophoblastic malignancies. In addition it was found that serial determinations of this protein could be useful in following the course of the disease and could aid in the evaluation of therapy (TATARINOV et al., 1974; SEPPÄLÄ et al., 1978; SEARLE et al., 1978).

Immunohistochemical detection of SP$_1$ in non-trophoblastic malignant tumours is another distinct possibility. HORNE et al., (1976a) could localize SP$_1$ and hPL (human placental lactogen) in the tumour cells of breast cancer patients. Positive results with SP$_1$ were found in 76% of cases, and with hPL in 82%. Cases negative for SP$_1$ and hPL showed significantly longer survival; SP$_1$ appeared to be a better indicator of prognosis than hPL.

The appearance of SP$_1$ in serum at or around the time of implantation of the human zygote suggests that it might be an appro-

Table 5.12. Procedure for isolation of PP_5

Placental fraction V	
DEAE-cellulose (batch)	
Sephadex G-150	
Immunoadsorption	
Removal of impurities	
(hPL, SP_1, IgG) with corresponding immunoadsorbents	
Yield	17%
Purity	> 99%

priate target for the immunological destruction of an early pregnancy. This possibility has already been investigated in female cynomolgus monkeys. An antifertility effect could be demonstrated in these animals by active or passive immunization with human SP_1 (BOHN and WEINMANN, 1974, 1976). Antibodies to SP_1 might also be used for immunotherapy of carcinomas producing this carcino-placental antigen.

5.2.2 Placental Protein PP_5

5.2.2.1 Occurrence

PP_5 is a soluble placental tissue protein which cannot be detected in other tissues (BOHN, 1972c). The average amount in a term placenta is around 1.5 mg, and a protein immunochemically identical to PP_5 can be detected in the placenta of apes and monkeys (BOHN and SEDLACEK, 1975). PP_5 is localized in the syncytiotrophoblast and the villous stroma of the term placenta (SEDLACEK et al., 1976). In the early placenta (8 weeks) it is localized almost exclusively in the syncytiotrophoblast (JONES, 1976). PP_5 is secreted into the maternal circulation where it can be detected by a sensitive radioimmunoassay.

5.2.2.2 Isolation

PP_5 has been isolated from an extract of human term placentae by an immunoadsorption technique (BOHN and WINCKLER, 1977a). Starting with placental fraction V the protein was first concentrated by adsorption to DEAE-cellulose and by gel filtration on Sephadex G-150. The eluate was applied to a PP_5-immunoadsorbent and eluted with 3 M potassium thiocyanate. The product still contained impurities (immunoglobulins, hPL and SP_1) which were removed by the use of appropriate antibodies coupled to CNBr-Activated Sepharose 4B. The purification scheme is shown in Table 5.12.

5.2.2.3 Characterization

The physical properties of PP_5 are summarized in Table 5.13. The molecular weight was determined by ultracentrifugation; the electrophoretic mobility was investigated in agar gel and on cellulose acetate strips.

PP_5 is a glycoprotein and contains 19.8% carbohydrate (Table 5.14). The most abundant amino acids are aspartic acid, glutamic acid, leucine and alanine, and a relatively high content of hemicystine (Table 5.15).

The biological function of PP_5 may be protease inhibition. Studies with fibrin agar electrophoresis plates revealed that the protein inhibits the proteolytic activity of trypsin and plasmin (BOHN and WINCKLER, 1977a).

Table 5.13. Physical properties of PP_5

Sedimentation coefficient	2.8 S
Molecular weight	36,600
Isoelectric point	4.6
Extinction coefficient E $_{1\ cm}^{1\%}$ (280 nm)	10.6
Electrophoretic mobility	β_1

Table 5.14. Carbohydrate composition of PP_5 (mg/100 mg)

Hexoses	10.0
Hexosamines[a]	4.4
Fucose	0.4
Neuraminic acid[a]	5.0
	19.8

[a] AS N-acetyl derivates.

5.2.2.4 Quantitation and Clinical Significance

The diagnostic significance of PP_5 determinations is now under evaluation. PP_5 appears to be specific for the trophoblast and may also prove to be a carcino-placental antigen, though it remains to be investigated whether PP_5 is produced by trophoblastic and other tumours. Injection of rabbit antibodies to PP_5 can induce abortion in cynomolgus monkeys. A significant reduction in fertility also has been observed in female cynomolgus monkeys actively immunized with PP_5 (BOHN and WEINMANN, in preparation). The antifertility effect of immunization with PP_5 may find application in the development of contraceptive vaccines as a new approach to birth control in humans (HARPER, 1975).

Table 5.15. Amino acid composition of PP_5 (residues per 100 residues)

Lysine	5.10	Alanine	7.55
Histidine	1.06	Cystine/2	7.12
Arginine	6.02	Valine	4.40
Aspartic acid	12.36	Methionine	0.91
Threonine	5.40	Isoleucine	2.60
Serine	5.74	Leucine	7.75
Glutamic acid	9.66	Tyrosine	4.89
Proline	4.73	Phenylalanine	5.61
Glycine	7.13	Tryptophan	1.88

References

Bohn, H.: Kristallisation und Hochreinigung des humanen Plazenta-Laktogens. Experientia 27, 1223-1225 (1971a)

Bohn, H.: Nachweis und Charakterisierung von Schwangerschaftsproteinen in der menschlichen Plazenta, sowie ihre quantitative immunologische Bestimmung im Serum schwangerer Frauen. Arch. Gynäkol. 210, 440-457 (1971b)

Bohn, H.: Isolierung und Charakterisierung des schwangerschafts-spezifischen β_1-Glykoproteins. Blut 24, 292-302 (1972a)

Bohn, H.: Charakterisierung der schwangerschafts-assoziierten Glykoproteine als akute Phase-Proteine. Arch. Gynäkol. 213, 54-72 (1972b)

Bohn, H.: Nachweis und Charakterisierung von löslichen Antigenen in der menschlichen Plazenta. Arch. Gynäkol. 212, 165-175 (1972c)

Bohn, H.: Isolierung des Plazenta-Proteins PP_2 und seine Identifizierung als Ferritin. Arch. Gynäkol. 215, 263-275 (1973)

Bohn, H.: Untersuchungen über das schwangerschafts-spezifische β_1-Glykoprotein. Arch. Gynäkol. 216, 347-358 (1974a)

Bohn, H.: Immunochemical determination of human pregnancy proteins. Arch. Gynäkol. 217, 219-231 (1974b)

Bohn, H.: The protein antigens of human placenta as a basis for the development of contraceptive vaccine. Development of vaccines for fertility regulation, WHO session: Third international symposium of immunology of reproduction, Varna Bulgaria 21-25 September 1975, pp. 111-125, 1975

Bohn, H.: Isolierung und Charakterisierung eines hochmolekularen Plazenta-Proteins (PP_6). Arch. Gynäkol. 218, 131-142 (1975b)

Bohn, H.: Isolation and characterization of placental specific proteins SP_1 and PP_5. Protides of biological fluids 24th Colloquium 1976, pp. 117-123 1976

Bohn, H., Kraus, W.: Isolierung und Charakterisierung des Plazenta-Proteins PP_1. Arch. Gynäkol. 221, 73-81 (1976)

Bohn, H., Kraus, W.: Isolierung und Charakterisierung des schwangerschafts-spezifischen β_1-Glykoproteins aus dem Urin schwangerer Frauen. Arch. Gynäkol. 223, 33-39 (1977)

Bohn, H., Ronneberger, H.: Immunologischer Nachweis von Schwangerschaftsproteinen des Menschen im Serum trächtiger Tiere. Arch. Gynäkol. $\underline{215}$, 277-284 (1973)

Bohn, H., Schwick H.G.: Isolierung und Charakterisierung eines fibrinstabilisierenden Faktors aus menschlichen Plazenten. Arzneim. Forsch. $\underline{21}$, 1432-1439 (1971)

Bohn, H., Sedlacek, H.H.: Eine vergleichende Untersuchung von Plazenta-spezifischen Proteinen bei Mensch und subhumanen Primaten. Arch. Gynäkol. $\underline{220}$, 105-121 (1975)

Bohn, H., Weinmann, E.: Immunologische Unterbrechung der Schwangerschaft bei Affen mit Antikörpern gegen das menschlichen schwangerschafts-spezifische β_1-Glykoprotein (SP$_1$). Arch. Gynäkol. $\underline{217}$, 209-218 (1974)

Bohn, H., Weinmann, E.: Antifertilitätswirkung einer aktiven Immunisierung von Affen mit dem schwangerschafts-spezifischen β_1-Glykoprotein (SP$_1$) des Menschen. Arch. Gynäkol. $\underline{221}$, 305-312 (1976)

Bohn, H., Winckler, W.: Isolierung und Charakterisierung eines neuen Gewebsproteins (PP$_7$) aus menschlichen Plazenten. Arch. Gynäkol. $\underline{222}$, 5-13 (1977a)

Bohn, H., Winckler, W.: Isolierung und Charakterisierung des Plazenta-Proteins PP$_5$. Arch. Gynäkol. $\underline{223}$, 179-186 (1977)

Bohn, H., Haupt, H., Kranz, Th.: Die molekulare Struktur der fibrinstabilisierenden Faktoren des Menschen. Blut $\underline{25}$, 235-248 (1972)

Bohn, H., Schmidtberger, R., Zilg, H.: Isolierung des schwangerschafts-spezifischen β_1-glykoproteins (SP$_1$) und antigenverwandter Proteine durch Immunadsorption. Blut $\underline{32}$, 103-113 (1976)

Gordon, Y.B., Grudzinskas, J.G., Jeffrey, D., Chard, T.: Levels of pregnancy-specific β_1-glycoprotein (SP$_1$) in maternal blood in normal pregnancy and in intrauterine growth retardation. Lancet 1977I, 331-333

Grenner, G.: Enzymimmunoassay zur Bestimmung des schwangerschafts-spezifischen β_1-Glykoproteins (SP$_1$). Analytika '78 München, 17-20.4.1978

Grudzinskas, J.G., Gordon, Y.B., Jeffrey, D., Chard, T.: Specific and sensitive determination of pregnancy specific β_1-glycoprotein (SP$_1$) by radioimmunoassay: a new pregnancy test. Lancet 1977aI, 333-335

Grudzinskas, J.G., Lenton, E.A., Gordon, J.B., Kelso, I.M., Jeffrey, D., Sobowale, O., Chard, T.: Circulating levels of pregnancy-specific β_1-glycoprotein in early pregnancy. Brit. J. Obstet. Gynaecol. $\underline{84}$, 740-742 (1977b)

Grudzinskas, J.G., Gordon, Y.B., Al-Ani, A.T.M.: Pregnancy-specific beta$_1$-glycoprotein in plasma and tissue extract in malignant teratoma of the testis. Br. Med. J. 1977c, 951-952

Harper, M.J.K.: The aim of the force on immunological methods for fertility regulation within the framework of the WHO expanded programme of research, development and research training in human reproduction. Development of vaccines for fertility regulation. WHO session: Third international symposium on immunology of reproduction, Varna Bulgaria September 1975, pp. 21-25, 1975

Heyderman, E.: personal communication 1978

Horne, C.H.W., Milne, G.D., Reid, I.N.: Prognostic significance of inappropriate production of pregnancy proteins by breast cancers. Lancet 1976a II, 279

Horne, C.H.W., Towler, C.M., Pugh-Humphreys, R.P.G., Thompson, A.W., Bohn, H.: Pregnancy-specific β_1-glycoprotein a product of syncytiotrophoblast. Experientia 32, 1197 (1976b)

Jones, W.R.: personal communications 1976

Neri, P., Arezzini,C., Fruschelli, C., Mueller, E.E.,Fioretti, P., Genazzani, A.R.: Effects of human chorionic somatomammotropin on the male reproductive apparatus of rodents and on placental steroids during human pregnancy, Excerpta Med. Int. Congr. Ser. 1976, 381 (Growth Horm. Relat. Pept.), 345-368 (1976)

Searle, F., Leake, B.A., Bagshawe, K.D., Dent, J.: Serum SP_1-pregnancy-specific β-glycoprotein in choriocarcinoma and other neoplastic disease. Lancet 1978 I, 579-580

Sedlacek, H.H., Rehkopf, R., Bohn, H.: Immunofluorescence histological localization of human pregnancy and placenta proteins in the placenta of man and monkeys (cynomolgus). Behring Inst. Mitt. No. 59, 81-91 (1976)

Seppälä, M., Rutanen, E.M., Heikinheimo, M., Jalanko, H., Engvall, E.: Detection of trophoblastic tumour activity by pregnancy-specific β_1-glycoprotein. Int. J. Cancer 21, 265-267 (1978)

Stevens, V.C., Bohn, H., Powell, E.J.: Serum levels of a placental protein during gestation in the baboon. Am. J. Obstet. Gynecol. 124, 51-54 (1976)

Tatarinov, Yu. S., Sokolov, A.V.: Development of a radioimmunoassay for pregnancy-specific $beta_1$-globulin and its measurement in serum of patients with trophoblastic and non-trophoblastic tumours. Int. J. Cancer 19, 161-166 (1977)

Tatarinov, Yu. S., Mesnyankina, N.V., Nikoulina, D.M., Toloknov, B.O., Falaleeva, D.M.: Identification immunochimique de la $beta_1$ globuline de la "zone de grossesse" dans la serum de malades atteintes de tumeurs trophoblastiques. Int. J. Cancer 14, 548-554 (1974)

Tatarinov, Yu. S., Falaleeva, D.M., Kalashnikov, V.V., Toloknov, B.O.: Immunofluorescent localisation of human pregnancy-specific-β-globulin in placenta and chorioepithelioma. Nature 260, 263 (1976)

Tatra, G., Breitenecker, G, W.: Serum concentration of pregnancy-specific-β_1-glycoprotein (SP_1) in normal and pathologic pregnancies. Arch. Gynäkol. 217, 383-390 (1974)

Tatra, G., Polak, S., Placheta, P.: Konzentration des schwangerschaftsspezifischen Proteins SP-1 im Fruchtwasser bei normalen und pathologischen Schwangerschaften. Arch. Gynäkol. 221, 161-166 (1976)

Towler, C.M., Horne, C.H., Jandial, V., Campbell, D.M., MacGillivray, I.: Plasma levels of pregnancy-specific β_1-glycoprotein in normal pregnancy. Br. J. Obstet. Gynaecol. 83, 775-779 (1926)

Towler, C.M., Horne, C.H.W., Jandial, V., Campbell, D.M., MacGillivray, I.: Plasma levels of pregnancy-specific β_1-glycoprotein in complicated pregnancies. Br. J. Obstet. Gynaecol. 84, 258-263 (1977)

Würz, H.: personal communication 1978

6 Pregnancy-Associated Plasma Proteins: PAPP-A and PAPP-B[1]

S.P.Halbert and T.-M.Lin

During pregnancy, it has long been known that certain proteins appear in the plasma that are apparently absent from it in the non-pregnant state. As part of a systematic attempt to detect abnormal proteins in the circulation during several disease states, our laboratory first approached a study of this question by immunological methods in the early 1960s. It was reasoned that unique proteins in the patient's plasma might evoke antibody responses in experimental animals which could be rendered apparent by absorption of the antisera with normal plasma. One of the disease states investigated was toxemia of pregnancy, but it was soon found that normal pregnancy plasma gave equivalent results. Hyperimmune antisera were prepared in rabbits by repeated immunizations with pooled third-trimester pregnancy plasma. When the antisera had achieved adequate titre, they were exhaustively absorbed with non-pregnant female and/or male plasma. Satisfactory lots of the resulting reagent reproducibly revealed up to four proteins in late pregnancy plasma or serum by various gel diffusion tests; these were designated pregnancy-associated plasma proteins, PAPP-A, -B, -C, and -D (GALL and HALBERT, 1972; LIN et al., 1973; 1974b; 1974c). These proteins were undetectable in cord sera and in non-pregnancy plasma, male or female, the latter with or without hormonal contraception.

---

1 The investigations were supported by a research grant from the National Institutes of Health (HD-05736).

Of the four proteins thus detected in pregnancy plasma, two were subsequently identified with known pregnancy proteins (LIN et al., 1974c). PAPP-D was found to represent the hormone, human placental lactogen (hPL or hCS), while PAPP-C was revealed to be identical to the pregnancy-specific β_1-glycoprotein (PSβG or SP$_1$) independently characterized by BOHN (1971), and TATARINOV and MASYUKEVICH (1970). The remaining two, PAPP-A and PAPP-B were shown to be unrelated to the other plasma proteins currently known to be increased in concentration during gestation (GALL and HALBERT, 1972; LIN et al., 1974b,c, 1978a; LIN and HALBERT 1975). The present report summarizes the present state of knowledge concerning these newly described pregnancy-specific plasma proteins.

6.1 Identification of PAPP

The most satisfactory method for demonstrating the pregnancy-associated plasma proteins with absorbed polyvalent anti-PAPP sera was by crossed immunoelectrophoresis. With this method, four precipitin "hills" were regularly seen with late pregnancy plasma samples, while none were detected with non-pregnancy plasmas. These systems are shown in Fig. 6.1. Of the four, the first to appear was always the PSβG (PAPP-C) reaction. It was the most dense, showed a β_1 mobility, and was always readily visible without staining for protein. The second most dense reaction was the PAPP-A system, of α_2 mobility. It appeared a little later than PSβG, and was generally visible without staining. The third system to be seen was the one identified as hPL, also of α_2 mobility. It was usually represented as a very fine faint line and frequently required staining to be clearly visualized. The last and faintest system to appear, PAPP-B, showed the same mobility as PSβG, but it required several days of incubation to be discernible. Even then, it usually could only be visualized after staining, and because of its location was often apparently masked by the very dense PSβG reaction. Two-directional immunodiffusion usually revealed only three systems with the same reagents, presumably because the PAPP-B reaction was too faint to be visualized or was overlapped by the other more intense precipitin lines.

It was shown that all the PAPPs were unrelated to various other pregnancy-associated plasma proteins, etc., with the aid of mono-specific antisera or purified proteins obtained from several sources, as listed in Table 6.1. Also recorded is the fact that none of the PAPPs were detected in various non-pregnancy human tissue extracts, nor did they show certain enzymatic activities (GALL and HALBERT, 1972; LIN et al., 1973, 1974 b,c; 1978 a; LIN and HALBERT, 1975).

6.2 Purification of PAPP

Submission of pregnancy serum to various fractionation procedures revealed several methods for the separation of PAPP-A from PAPP-B. For example, salting out whole pregnancy serum with 30% saturated ammonium sulfate (1.2 M) resulted in a precipitate which con-

Table 6.1. Pregnancy-associated proteins (and others) found to be unrelated to PAPP-A and PAPP-B, as well as PSβG and hPL

---

1. Human chorionic gonadotrophin (hCG)

2. α fetoprotein (AFP)

3. Carcinoembryonic antigen (CEA)

4. Pregnancy zone protein (PZP; PAAG) (same as α_2 pregnoglobulin; pregnancy-associated α_2 glycoprotein; new serum pregnancy α_2 macroglobulin; PAG; pregnancy-associated α_2 globulin; PA 1; Xh; and Xm proteins)

5. Placental alkaline phosphatase

6. Histaminase

7. Oxytocinase

8. Oxytocin

9. C-reactive protein

10. ρ-antigen (KNOX et al., 1975)

11. Placental proteins; PP1, PP2, PP3, PP4, and PP5 (BOHN, 1972a)

12. Sex-steroid-binding globulin (steroid-binding β-globulin) (BOHN and KRANZ, 1973; MERCIER-BODARD et al., 1970)

13. Lactoferrin

14. Prolactin (Bovine and ovine)

15. PAPP-A, PAPP-B, hPL and PSβG were not detected in saline extracts of non-pregnancy liver, kidney, pancreas, skeletal muscle or heart

16. PAPP-A, PAPP-B, hPL and PSβG were found to be devoid of the following enzyme activities by histochemical methods applied to the immune precipitates; acid phosphatase, glucuronidase, glucosaminidase, esterase and catalase

---

tained virtually all the PAPP-B and about one-half of the PSβG, while all the PAPP-A remained in solution in the supernate. Salting out the latter with 5% saturated ammonium sulfate (2.0 M) insolubulized the PAPP-A, as well as the remainder of PSβG and the hPL. Both PAPP-A and PAPP-B were essentially insoluble in distilled water.

Chromatography of term pregnancy serum on hydroxylapatite resulted in considerable separation of PAPP-A and PAPP-B (Fig. 6.2). The bulk of the PAPP-A eluted with the void volume (along with the PSβG and hPL), while the PAPP-B desorbed under slightly higher phosphate concentration, about 0.1 M. A subfraction with PAPP-A activity eluted at high salt concentration.

Chromatography of pregnancy serum on DEAE cellulose with stepwise elution resulted in partial separation of PAPP-A from PAPP-B (Table 6.2). In this case, considerable separation of these pregnancy proteins from PSβG and hPL was achieved (LIN et al., 1974b; 1978a).

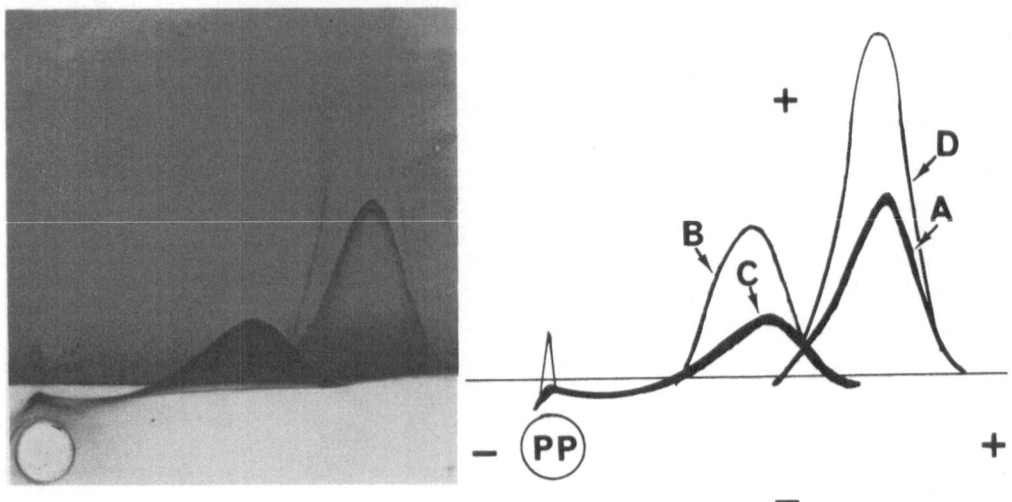

Fig. 6.1. Crossed immunoelectrophoresis of pregnancy plasma (*PP*) with polyvalent rabbit anti-PAPPs in the agarose gel (2nd direction). *A, B, C, D,*:
the immunoprecipitin "hills" of PAPP-A, PAPP-B, PSβG and hPL respectively.
+ and -, anode and cathode respectively for first and second directions

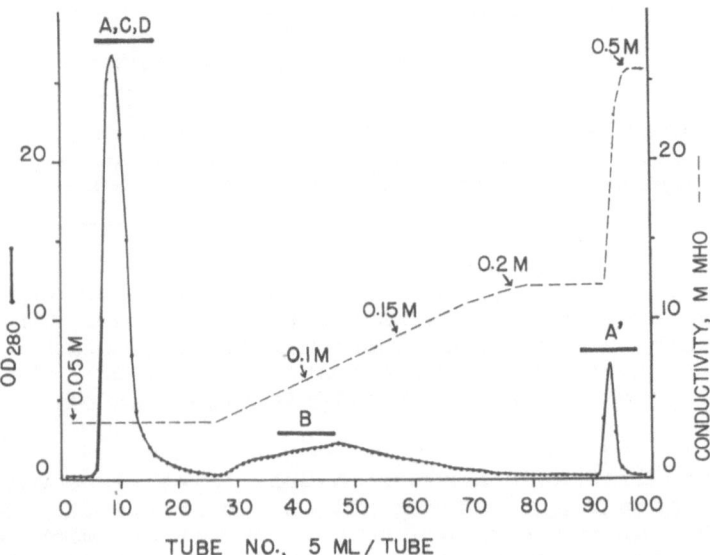

Fig. 6.2. Chromatography of term pregnancy serum on hydroxylapatite. Elution was carried out with a gradient of increasing potassium phosphate concentrations (pH 6.8). Representative concentrations are shown with arrows.
The horizontal bars show the fractions containing PAPP-A, PAPP-B, PSβG and
hPL (*A, B, C,* and *D* respectively). A' is a subfraction with PAPP-A activity

Table 6.2. DEAE-cellulose column chromatography of pregnancy serum[a]

Proteins	Eluting salt (_M_ NaCl) concentration				
	0.025	0.075	0.125	0.175	0.3
PAPP-A	-	-	-	++	++
PAPP-B	-	-	-	+++	+
PSβG	+	+++	-	-	-
hPL	-	-	+++	-	-

[a] 400 ml serum collected on the day of delivery was applied into a 8.2 cm (dia) x 11 cm (height) column in 0.025 _M_ NaCl with 0.005 _M_ sodium phosphate buffer at pH 7.5. It was eluted stepwise with increasing NaCl concentrations in the phosphate buffer.
-, not detectable or trace; +, ++, +++, increasing concentrations of each PAPP.

Both PAPP-A and PAPP-B were excluded by Sephadex G-200 gel filtration, in contrast to hPL and PSβG. In addition, PAPP-A and PAPP-B could be separated from each other in pregnancy serum by agarose gel filtration. This is illustrated in Fig. 6.3, utilizing Sepharose 4B. It may be seen that under these conditions, PAPP-B eluted shortly after the void volume, followed by PAPP-A. The PZP eluted just behind PAPP-A, and well before PSβG and hPL. The PAPP-B eluted almost simultaneously with immunoglobulin M. Other similar analyses on calibrated Sepharose 6B and Sepharose 4B columns indicated that PAPP-B has a molecular size of about 1,000,000, while PAPP-A was about 750,000. Some discrepancies between these molecular weight estimates and preliminary ultracentrifugation analyses were seen, but it is known that such values are affected by a number of other variables, e.g. molecular shape, chemical composition, etc. (LIN et al., 1978a).

Based on these observations, PAPP-A and PAPP-B were each purified to a considerable degree from term pregnancy serum. The steps utilized for the isolation of PAPP-B are summarized in the flow diagram of Fig. 6.4. It can be seen that the final fraction obtained from 1.2 litres of pregnancy serum was only about 5 mg, but that its specific immunological reactivity was about 800 times greater than the starting material. It could be roughly estimated that the yield was about 10%. Similar preparations of purified PAPP-B were shown to be clearly reactive with polyvalent anti-PAPP sera in immunodiffusion, as shown in Fig. 6.5. The reactions of non-identity of the PAPP-B system with PAPP-A, PSβG and hPL are evident.

PAPP-A was purified from pregnancy serum according to the flow diagram of Fig. 6.6. It may be seen that the final preparation was 115-fold purified over the starting material, although it was still contaminated with small amounts of some normal serum proteins. Other separation methods substituting for the last gel filtration step of Fig. 6.6 proved less satisfactory. These included isoelectric focussing and column gradient solubilization with ammonium sulfate. Immunodiffusion reactions with

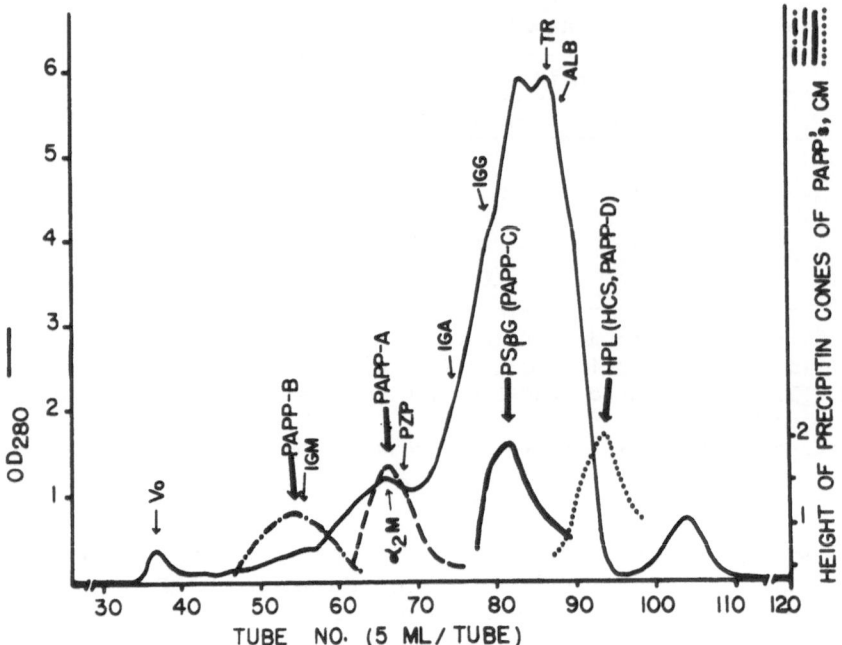

Fig. 6.3. Gel filtration of term pregnancy serum on Sepharose 4B. The heights of the precipitin "rocket" cones are recorded for PAPP-B (-.-); PAPP-A (----); and hPL (....). The *continuous solid line* represents the absorbance at 280 nm. The *arrows* indicate the tubes at which the various normal serum proteins peaked, as estimated by immunoassay: immunoglobulin M, A and G (*IgM, IgA, IgG*); α_2 macroglobulin (α_2M); pregnancy zone protein (*PZP*); transferrin (*TR*); and albumin (*ALB*). Vo, void volume

purified PAPP-A and the polyvalent anti-PAPPs are shown in Fig. 6.7. By use of appropriate stains on the immune precipitates, it was shown that PAPP-A and PAPP-B were devoid of lipids (LIN et al., 1974c; 1978a).

6.3 Physicochemical Characterization of PAPP

Both PAPP-A and PAPP-B appear to be glycoproteins containing sialic acid, since exposure of term pregnancy serum to several neuraminidase preparations resulted in an appreciable loss of electrophoretic mobility (LIN et al., 1974c; 1978a). This loss was greater for PAPP-A than PAPP-B, but neither was affected as much as PSβG (PAPP-C), which is known to contain large amounts of sialic acid (BOHN, 1972b). These findings may suggest that the percent content of sialic acid in these proteins decreases in the following order: PSβG, PAPP-A and PAPP-B. Purification of each of the latter two proteins to unequivocal homogeneity will be required to establish this fact on a chemical basis. Both PAPP-A and PAPP-B were inactivated by treatment with trypsin, but were unaffected by desoxyribonuclease or ribonuclease.

Term pregnancy serum (1200 ml) (arbitrarily assigned 1 unit PAPP-B/mg protein)

⇓

Collect ppt. after salting out with 30% saturated ammonium sulfate (1.2 M)

⇓

Dialyze vs 0.005 M sodium phosphate, pH 7.5
Chromatography on DEAE cellulose with stepwise elution
Harvest and pool protein peaks eluting at 0.175 M and 0.3 M NaCl

⇓

Collect ppt. after salting out with 30% saturated $(NH_4)_2SO_4$
Dialyze vs PBS

⇓

Subject to Sepharose 4B gel filtration. Harvest early fractions rich in PAPP-B

⇓

Collect ppt. after salting out with 30% saturated $(NH_4)_2SO_4$
Dialyze vs 0.05 M potassium phosphate, pH 6.8

⇓

Apply to hydroxylapatite column. Flush with starting buffer.

⇓

Elute PAPP-B with 0.15 M potassium phosphate, pH 6.8

⇓

Collect ppt. after salting out with 30% saturated ammonium sulfate
Dialyze vs PBS, store -20°

⇓

(Final yield 5 mg, with specific immunological reactivity about 800X over starting pregnancy serum)

Fig. 6.4. Flow diagram for purification of PAPP-B from term pregnancy serum

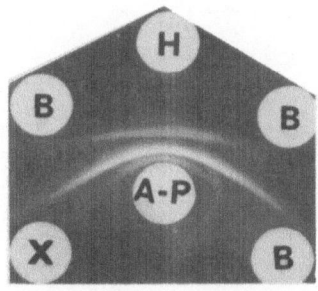 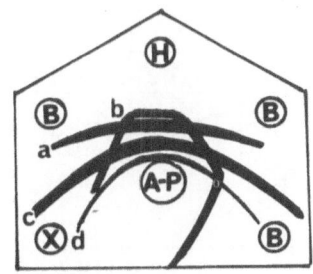

Fig. 6.5. Immunodiffusion reactions between polyvalent anti-PAPP serum *(A-P)* and purified PAPP-B *(B)* at 1 mg/ml. *H* term pregnancy serum. *X* phosphate buffered saline (PBS). *a*, *b*, *c* and *d* represent the precipitin lines of the PAPP-A, PAPP-B, PS*B*G and hPL systems respectively

Term pregnancy plasma (400 ml), dialyzed vs 0.005 M sodium phosphate pH 7.5

DEAE cellulose chromatography with $NaCl_2$ gradient elution. PAPP-A eluted between 0.175 and 0.3 M NaCl.

Dialyze vs PBS, Sephadex G-200 gel filtration. Void volume contained PAPP-A.

Sepharose 6B gel filtration. Pool PAPP-A peak.

Salt out with ammonium sulfate (1/3 to 2/3 saturation step).

Harvest precipitate, dialyze in PBS, store at -20°C.

(This fraction was purified about 115-fold in terms of immunological re-activity)

Fig. 6.6. Flow diagram for purification of PAPP-A from late pregnancy serum

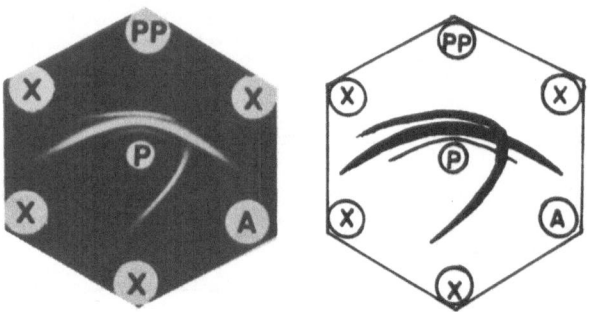

Fig. 6.7. Immunodiffusion identification of the purified PAPP-A with the polyvalent anti-PAPPs (P) third trimester pregnancy plasma; A, purified PAPP-A at 0.5 mg/ml; X, buffered saline

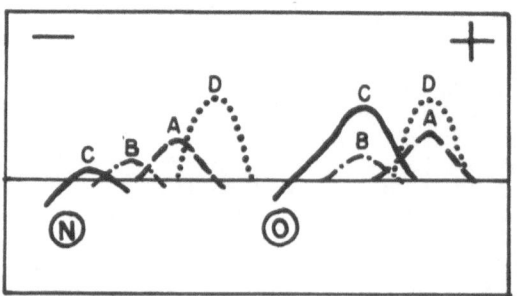

Fig. 6.8. Crossed Immunoelectrophoresis of the pregnancy proteins in term pregnancy serum after treatment with neuraminidase (N) or with buffer (O). The agar gel layer of the second direction contained polyvalent anti-PAPPs. A, B, C and D refer to the immune precipitin systems due to PAPP-A, PAPP-B, PSβG, and hPL respectively

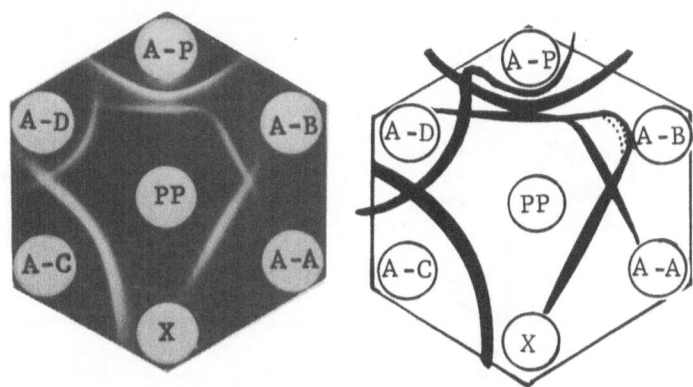

Fig. 6.9. Immunodiffusion reactions with monospecific antiserum to PAPP-A
(*A-A*) and PAPP-B (*A-B*), as well as anti-PSβG (*A-C*) and anti-hPL (*A-D*), test-
ed against term pregnancy serum supplemented with purified PAPP-B (*PP*). PBS,
X. Polyvalent anti PAPPs (*A-P*)

6.4 Immunological Characterization of PAPP

Monospecific antisera to PAPP-A and PAPP-B were obtained by repeated immunization of rabbits with partially purified preparations in Freund's adjuvant. Both had to be absorbed with non-
pregnancy human plasma and/or plasma of women taking estrogen
contraceptives to achieve monospecifity. In the case of the
PAPP-B antiserum, sometimes small amounts of contaminating anti-
PAPP-A and anti-PSβG were removed by absorption with the 30%-
50% saturated ammonium sulfate precipitate obtained from term
pregnancy serum. As indicated above, this fraction was essentially devoid of PAPP-B, but contained the other pregnancy proteins. The reaction of the monospecific antisera to PAPP-A and
PAPP-B with term pregnancy serum are shown in Fig. 6.9. More
definitive identification of the monospecific antisera for the
PAPP-A and PAPP-B systems was made by the intermediate gel
crossed immunoelectrophoretic method (AXELSEN et al., 1973;
LIN et al., 1978a). Monospecific antisera placed in the intermediate gels eliminated the appropriate reaction and confirmed
the identity of each system with the "hills" found in the
crossed immunoelectrophoretic pattern of Fig. 6.1. These findings
are illustrated in Fig. 6.10, which also confirms the identity
of the PSβG and hPL reactions.

Other primate species were found to possess pregnancy-specific
proteins cross-reactive with antibodies to the human PAPPs, and
these proteins were not seen in non-pregnancy primate serum (LIN
and HALBERT, 1978). In the case of the pregnant chimpanzee and
orangutan, their PAPP-A and PAPP-B showed apparent immunological
reactions of identity with the human proteins, while in Old
World monkeys partial identity patterns were seen. Similar patterns of cross-reactivities were seen for PSβG and hPL analogs
in these primate species. The pregnant New World (squirrel) monkey did not reveal any cross-reactive PAPP-A or PAPP-B.
None of the antibodies to the human pregnancy proteins cross-
reacted with pregnancy serum of the mouse or rat. However, hy-

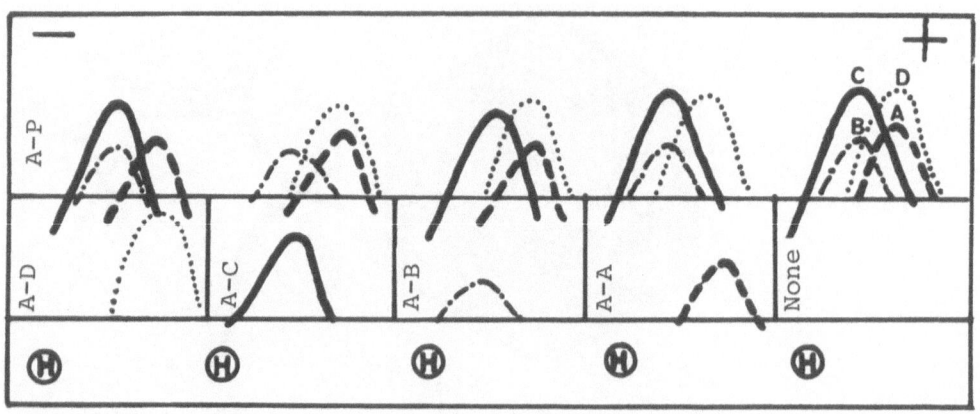

Fig. 6.10. Intermediate gel crossed-immunoelectrophoresis with monospecific anti-PAPPs, to identify the "hills" seen with the reference polyvalent anti-PAPP antiserum (A-P), contained in the top gel compartment. H, term pregnancy serum. The intermediate gels contained anti-hPL (A-D; anti-PSβG (A-C); anti-PAPP-B (A-B); and anti-PAPP-A (A-A); and no antiserum (none). The immune precipitin systems were identified as A, B, C, and D representing PAPP-A, PAPP-B, PSβG and hPL respectively

perimmune antiserum against late pregnant rat serum exhaustively absorbed with non-pregnancy rat serum also revealed four rat pregnancy-associated plasma proteins. These showed somewhat similar patterns in crossed-immunoelectrophoresis as the human PAPPs, and the rat pregnancy proteins also were not seen in non-pregnancy rat serum. As expected, the polyvalent antisera to rat PAPPs did not cross-react with the human PAPPs. It was of interest that none of the rat PAPPs were seen in the serum of rats made pseudopregnant by electrical and mechanical stimulation, as shown in Fig. 6.11 (LIN et al., 1974a; 1978b).

6.5 Synthesis of PAPP

It is probable that both PAPP-A and PAPP-B are synthesized in the placenta. In addition to the fact that both appear to be specific for pregnancy plasma, both disappeared quite rapidly during the post-partum period as described below. In order to analyze the possible placental origins of these proteins more critically, detailed quantitative analyses of the PAPP contents of placental extracts were carried out (LIN et al., 1976a). Placental tissue fragments were extensively washed as free of blood as possible, homogenized in a blender, and the extracts were clarified by high-speed centrifugation. These extracts were assayed for their content of each PAPP, and the values were compared with those seen in the appropriate maternal serum. In addition, since the placental extracts unavoidably contained some maternal and fetal serum, the amounts were determined by parallel quantitation of several normal plasma proteins in the extracts and in maternal serum (IgA and IgM) principally maternal-blood derived; albumin and transferrin maternal- and fetal-

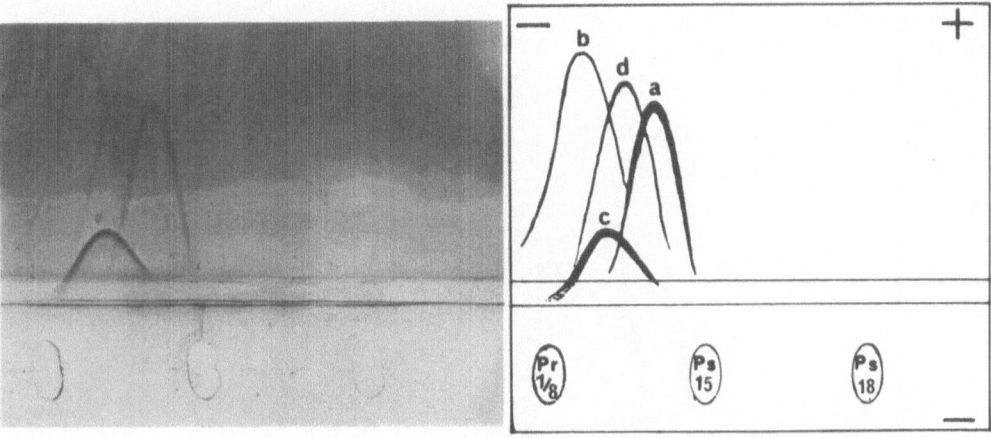

Fig. 6.11. Crossed immunoelectrophoresis of rat PAPPs and their absence in pseudo-pregnancy rat serum. The second gel contained polyvalent anti-rat PAPPs. *Pr 1/8*, late rat pregnancy serum diluted 1:8; *Ps 15* and *Ps 18*, pools of undiluted pseudo-pregnancy serum equivalent to 15 and 18 days of true pregnancy respectively (average term pregnancy was 21 days). All four reactions were due to the pregnancy serum; no reactions were seen with the pseudo-pregnancy sera. *a*, *b*, *c*, and *d*: rat pregnancy-associated plasma proteins, possibly analogous to human PAPP-A, PAPP-B, PSβG and hPL respectively

blood derived). The maternal serum contribution of each pregnancy protein in the placental extracts could thus be estimated. Such analyses demonstrated that all the PAPPs were present in placental extracts in significantly larger amounts than could be accounted for by their content of maternal serum. As expected, by far the greatest amount was found for hPL, followed by PAPP-B, PSβG and then PAPP-A. In sharp contrast, all of the pregnancy zone protein (PZP) detected in the placental extracts could be accounted for on the basis of their content of maternal blood.

Immunofluorescence studies have been carried out thus far on frozen sections of placenta with anti-PAPP-A. The results demonstrated localization of PAPP-A almost exclusively and evenly in the cytoplasm of the villous trophoblast, The localization was essentially the same as that seen for PSβG and hPL (LIN et al., 1975; LIN and HALBERT, 1976).

6.6 Physiology of PAPP

Analysis of the changes in plasma concentrations of PAPP-A and PAPP-B during gestation and the postpartum period demonstrated that they both rose slowly during the second trimester and more rapidly during the third trimester. PAPP-B tended to plateau in late pregnancy, while PAPP-A appeared to rise steadily until delivery (Fig. 6.12) (LIN et al., 1974d, 1978a). PAPP-B disappeared quite rapidly during the postpartum period, with an apparent half-life of less than 1 day, while PAPP-A revealed a half-life of roughly 3-4 days (LIN et al., 1976c). Also compared

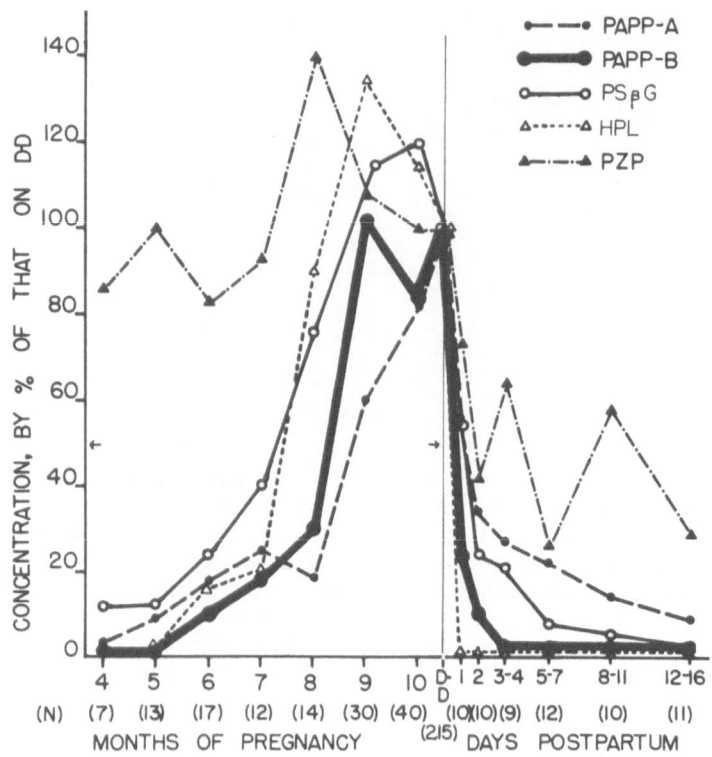

Fig. 6.12. Pregnancy protein concentrations in maternal plasma during gestation and the postpartum period. Pools of pregnancy plasma were made for each period studied; the number of specimens (N) in each pool is indicated in parenthesis. The values are recorded as a percentage of those found in a reference term pregnancy plasma pool, arbitrarily assigned a value of 100 for each pregnancy protein. *Arrow*, half-life following delivery

in Fig. 6.12 are the changes in the pregnancy proteins, PSβG, hPL and PZP. The half-life of PSβG was 1-2 days (BOHN, 1974; LIN et al., 1974d; 1976c), while the kinetics of PZP were quite distinct from the other four. It should be noted here that the PZP analogue in some Old World monkey species showed a decrease in concentration during pregnancy (LIN et al., 1976b).

The physicochemical properties and kinetic behaviour of these five pregnancy proteins are summarized in Table 6.3. It is probably that the peak absolute concentrations of PAPP-A and PAPP-B are lower than those of PSβG and PZP.

Pregnancy sera were obtained from normal women within 7 days of delivery of a single viable normal baby, and the relations of PAPP-A and PAPP-B concentrations to various obstetric parameters were examined (LIN et al., 1976d, 1978c). The PAPP-A level was found to be significantly correlated with placental weight and newborn weight, and possibly with diastolic blood pressure

Table 6.3. Properties of five human pregnancy-associated plasma proteins detectable by gel diffusion

Protein	IEP[c] mobility	Estimated molecular weight	pI[d]	Post-partum half-life	Peak in gestation (μg/ml)
PAPP-A	α_2	750,000	4.4	3-4 days	Very late third trimester (?)
PAPP-B	β_1	1,000,000	4.6-5.0	<1 day	Third trimester (?)
PSβG[a]	β_1	110,000	3.8	1-2 days	Late third trimester (250)
hPL[b]	α_2	20,000	5.7	20 mins.	Mid-third trimester (7)
Pregnancy zone (PZP)	α_2	300,000-506,000	4.4-4.8	1-3 weeks	Third trimester (850)

[a] Pregnancy specific β_1 glycoprotein (SP$_1$) (PAPP-C).

[b] Human placental lactogen, human chorionic somatomammotrophin (hCS) (PAPP-D).

[c] IEP, immunoelectrophoretic mobility.

[d] pI, isoelectric point.

and parity. No correlations were found with other parameters, including maternal weight and age, 1- and 5-min Agpar scores, systolic blood pressure, day of delivery, sex of the newborn, newborn bilirubin and weeks of gestation. PAPP-B was correlated with placental and newborn weights, but not with the other parameters.

6.7 PAPP in Pathological States

Preliminary investigations of the PAPP-A and PAPP-B levels in abnormal pregnancies have been carried out with plasma samples taken during the last month of gestation (LIN et al., 1977, 1978c). It was found that PAPP-A was significantly increased in toxemia (diastolic blood pressure greater than 110), while PAPP-B was decreased. In pregnancies complicated with diabetes (insulin required), PAPP-A was not affected, while PAPP-B was decreased. In diabetes of pregnancy and in toxemia, it was found that PSβG and hPL were within normal limits. In twin pregnancies, PAPP-A was strikingly elevated, while PAPP-B was somewhat increased, although not significantly so. PSβG and hPL also were increased in twin pregnancies. It is obvious that more extensive analysis of PAPP-A and PAPP-B in normal and abnormal pregnancies are indicated. Such studies would best be carried out with more sensitive and precise assays for measuring these proteins, such as radioisotope- or enzymelabeled immunoassays as described for PSβG (GRUDZINSKAS et al., 1976; TOWLER et al., 1977) and PZP (STIMSON and SINCLAIR, 1974). It is also clear that investigations aimed at discovering the function of these proteins would be of importance.

References

Axelsen, N.H., Kroll, J., Weeke, B.: A manual of quantitative immunoelectrophoresis. Methods and applications. Scand. J. Immunol. 2 (Suppl. 1), (1973)

Bohn, H.: Nachweis und Charakterisierung von Schwangerschaftsproteinen in
der menschlichen Plazenta, sowie ihre quantitative immunologische Bestimmung im Serum schwangerer Frauen. Arch. Gynaekol. 210, 440-457 (1971)

Bohn, H.: Nachweis und Charakterisierung von löslichen Antigenen in der
menschlichen Plazenta. Arch. Gynaekol. 212, 165-175 (1972a)

Bohn, H.: Isolierung und Charakterisierung des schwangerschafts-spezifischen
β_1-Glykoproteins. Blut 24, 292-302 (1972b)

Bohn, H.: Untersuchungen über das schwangerschafts-spezifische β_1-Glykopro-
tein (SP$_1$). Arch. Gynaekol. 216, 347-358 (1974)

Bohn, H., Kranz, T.: Untersuchungen über die Bindung von Steroidhormonen an
menschliche Schwangerschaftsproteine. I. Identifizierung des schwangerschafts-assoziierten β_1-Glykoproteins mit dem Steroid-bindenden β-Globulin.
Arch. Gynaekol. 215, 63-71 (1973)

Gall, S.A., Halbert, S.P.: Antigenic constituents in pregnancy plasma which
are undetectable in normal, non-pregnant female or male plasma. Int. Arch.
Allergy Appl. Immunol. 42, 503-515 (1972)

Grudzinskas, J.G., Gordon, Y.B., Jeffrey, D., Chard, T.: Specific and sensitive determination of pregnancy-specific β_1-glycoprotein by radioimmunoassay. Lancet 1976 I, 333-335

Knox, E., Andiman, W.A., Cappel, R., Schluederberg, A., Hobbins, J.C., Horst-
man, D.M.: The relationship of a newly described acute phase protein to
human gestation. Am. J. Obstet. Gynecol. 122, 955-957 (1975)

Lin, T.M., Halbert S.P.: Immunological comparison of various human pregnancy-
associated plasma proteins. Int. Arch. Allergy Appl. Immunol. 48, 101-115
(1975)

Lin, T.M., Halbert, S.P.: Placental localization of human pregnancy-associated
plasma proteins. Science 193, 1249-1252 (1976)

Lin, T.M., Halbert, S.P.: Immunological relationship of human and subhuman
primate pregnancy-associated plasma proteins. Int. Arch. Allergy Appl.
Immunol. 56, 207-223 (1978)

Lin, T.M., Halbert, S.P., Kiefer, D.: Characterization and purification of
human pregnancy-associated plasma proteins. Fed. Proc. 32, 623 (1973)

Lin, T.M., Halbert, S.P., Kiefer, D.: Pregnancy-associated serum antigens
in the rat and mouse. Proc. Soc. Exp. Biol. Med. 145, 62-66 (1974a)

Lin, T.M., Halbert, S.P., Kiefer, D., Spellacy, W.N. Gall, S.: Characterization of four human pregnancy-associated plasma proteins. Am. J. Obstet.
Gynecol. 118, 223-236 (1974b)

Lin, T.M., Halbert, S.P., Kiefer, D., Spellacy, W.N.: Three pregnancy-associated human plasma proteins. Purification, monospecific antiserum and
immunological identification. Int. Arch. Allergy Appl. Immunol. 47, 35-53
(1974c)

Lin, T.M., Halbert, S.P., Spellacy W.N.: Measurements of pregnancy-associated
plasma proteins during human gestation. J. Clin. Invest. 54, 576-582 (1974d)

Lin, T.M., Halbert, S.P., Spellacy, W.N., Gall, S.: Measurement of pregnancy-associated plasma proteins (PAPPs) during gestation and their immunological identification. Fed. Proc. 33, 282 (1974e)

Lin, T., Halbert, S.P., Kiefer, D.: Pregnancy-associated plasma proteins (PAPP's) in human placenta. Fed. Proc. 34, 682 (1975)

Lin, T.M., Halbert, S.P., Kiefer, D.: Quantitative analysis of pregnancy-associated plasma proteins in human placenta. J. Clin. Invest. 57, 466-472 (1976a)

Lin, T.M., Halbert, S.P., Plasencia, R.: Pregnancy zone protein analogue in pregnant and non-pregnant primates, and its decrease during pregnancy in some monkey species. Clin. Exp. Immunol. 26, 609-622 (1976b)

Lin, T.M., Halbert, S.P., Spellacy, W.N., Gall, S.: Human pregnancy-associated plasma proteins during the post-partum period. Am. J. Obstet. Gynecol. 124, 382-387 (1976c)

Lin, T.M., Halbert, S.P., Spellacy, W.N.: Relation of obstetric parameters to the concentrations of four pregnancy-associated plasma proteins at term in normal gestation. Am. J. Obstet. Gynecol. 125, 17-24 (1976d)

Lin, T.M., Halbert, S.P., Spellacy, W.N., Berne, B.H.: Plasma concentrations of four pregnancy proteins in complications of pregnancy. Am. J. Obstet. Gynecol. 128, 808-810 (1977)

Lin, T.M., Halbert, S.P., Kiefer, D.: Characterization and purification of pregnancy-associated plasma protein B (PAPP-B). Int. Arch. Allergy Appl. Immunol. (1978a) (in press)

Lin, T.H., Sperling, F., Lin, T.M., Halbert, S.P.: Pregnancy-associated plasma proteins in pregnancy and pseudopregnant rats. Int. Arch. Allergy Appl. Immunol. (1978b) (in press)

Lin, T.M., Halbert, S.P., Spellacy, W.N.: Pregnancy-associated plasma protein B (PAPP-B) in normal and abnormal pregnancies at term. Br. J. Obstet. Gynaecol. (1978c) (in press)

Mercier-Bodard, C., Alfsen, A., Baulieu, E.E.: Sex steroid binding plasma protein (SBP). Karolinska symposium on research methods of reproductive endocrinology, second symposium pp. 204-224, 1970

Stimson, W.H., Sinclair, J.M.: An immunoassay for a pregnancy-associated α-macroglobulin using antibody enzyme conjugates. FEBS Lett. 47, 190-192 (1974)

Tatarinov, Y.S., Masyukevich, V.N.: Immunochemical identification of new beta$_1$-globulin in the blood serum of pregnant women. Biull. Eksp. Biol. Med. 69, 66-68 (1970)

Towler, C.M., Horne, C.H.W., Jandial, V., Chesworth, J.M.: A simple and sensitive radioimmunoassay for pregnancy-specific-β_1-glycoprotein. Br. J. Obstet. Gynaecol. 84, 580-584 (1977)

7 Observations on the Isolation of Pregnancy-Associated Plasma Protein A[1]

P. Bischof [2]

Up to four antigenic constituents have been found in the plasma of pregnant women which are not detectable in non-pregnant female or male plasma (GALL and HALBERT, 1972). These antigens have been named "pregnancy-associated plasma proteins" (PAPP) and lettered "A, B, C, D" according to the position of their precipitin line on an immunodiffusion plate, starting from the antigen well.

Direct immunodiffusion has been used to identify these antigens and to compare them with proteins already known (LIN and HALBERT, 1975). These experiments showed an immunological identity between PAPP-C and SP_1 (Schwangerschaft spezifisches Protein 1, BOHN, 1972) and between PAPP-D and hPL (human placental lacto-

1 These investigations were supported by a research grant from the Royal
  Society.
2 The author is deeply indebted to Dr. N.A. Booth, to Dr. J.E. Fothergill,
  and to Professor A. Klopper, for their advice and interest in this work.

gen). None of the many pregnancy proteins studied showed identity
with PAPP-A or PAPP-B, so that these two antigens can be considered as new pregnancy proteins.

In this rapidly moving field, one and possibly two more pregnancy proteins were recently described (BOHN and WINCKLER,
1977a, b). Partial purifications and characterizations have
been published for PAPP-A and PAPP-B (LIN et al., 1974a, 1978).

PAPP-A was shown to be a large glycoprotein with a molecular
weight of 750,000, with an α_2 electrophoretic mobility and a
pI between 4.1 and 4.5. PAPP-B is an even larger $\beta_1$glycoprotein
with a molecular weight of 1,000,000. Clinical investigations
have been undertaken in which PAPP-A was measured by crossed
immunoelectrophoresis (LIN et al., 1974b) and PAPP-B by rocket
immunoelectrophoresis (LIN et al., 1978). It was shown that the
concentration of both proteins in the maternal circulation
increases as pregnancy progresses and rapidly decreases after
delivery with a half-life of 2-3 days for PAPP-A and less than
1 day for PAPP-B. Moreover, only trace amounts of both antigens
have been found in the fetal circulation.

The probable placental origin of PAPP-A and PAPP-B (LIN et al.,
1976, 1978) plus the fact that they are mainly secreted into the
maternal circulation, makes them likely candidates for the functional assessment of the fetoplacental unit. In order to test
this possibility, and because any investigation in this field
depends upon the availability of a reasonably pure preparation,
we started to isolate these antigens. The present report will
essentially deal with the purification and the method of measurement of PAPP-A.

7.1 Isolation of PAPP-A from Pregnancy Plasma

Purifying a protein from a biological material not only needs
techniques to separate it from other contaminating materials,
but also requires a reliable method for tracing the protein
throughout the purification procedure. The best purification
method will only be as good as the tracing method available.
The method we chose was to identify PAPP-A with a monospecific
antiserum obtained from Dr. Lin. This was done on simple immunodiffusion plates as described elsewhere (HUDSON and HAY, 1976).

In addition to an identification method, techniques for assessing the degree of purity are also required and this was performed by polyacrylamide gel electrophoresis (WEBER et al.,
1972; DAVIES, 1964).

The pioneering work of Lin has already defined some of the biochemical characteristics of PAPP-A. Knowing these was a great
help in designing the purification process.

7.1.1 Precipitation with Ammonium Sulphate

Precipitation with ammonium sulphate is usually the first step because it eliminates any unwanted proteins which remain soluble.

In practice, solid ammonium sulphate was slowly added to 1100 ml pooled third-trimester pregnancy plasma, in order to provide a 60% saturated solution. After centrifugation, the precipitate was redissolved in buffer A (0.005 M sodium phosphate pH 7.4). This solution was then thoroughly dialysed against buffer A in order to eliminate the unwanted ammonium sulphate.

7.1.2 Ion-Exchange Chromatography

PAPP-A has a pI of around 4.1 and therefore has a negative net charge at a pH of 7.4 (buffer A), and will bind to a matrix carrying positive charges such as DEAE-cellulose. This material was used as a 5x40 cm column equilibrated with buffer A. The redissolved precipitate was applied to the column and the bound proteins slowly eluted with 20 litres of a NaCl gradient (from 0 to 0.3 M) in buffer A.

The presence of proteins in the eluted fractions was determined by measuring the optical density at 280 nm. The presence of PAPP-A was assessed by submitting each of the 1300 fractions to immunodiffusion against PAPP-A antiserum. The linearity of the gradient was checked by measuring the conductivity of the eluate. The results obtained are shown in Fig. 7.1.

At this early stage in the purification, PAPP-A is already clearly separated from $PS\beta_1G$, (SP_1). It is eluted well after the bulk contaminating proteins, between 0.1 M and 0.2 M NaCl.

The PAPP-A containing fractions were pooled and precipitated with ammonium sulphate. The precipitate was redissolved in buffer B (0.05 M sodium phosphate pH 6.5 containing 0.08 M NaCl) and dialysed against this buffer.

Because the concentration of unwanted proteins was still high and because of the large volumes involved a further ion-exchange chromatography was performed, this time on DEAE-Sephadex. This was packed as a 5x10 cm column and equilibrated in buffer B. The material from DEAE-cellulose was applied to the DEAE-Sephadex column and elution carried out with 4 litres of a NaCl gradient (from 0.06 to 0.6 M) in buffer B. Again, each fraction was tested for optical density, presence of PAPP-A and conductivity. The elution profile obtained is shown in Fig. 7.2.

At this lower pH, PAPP-A is more tightly bound to the DEAE-Sephadex matrix and is eluted with about twice the concentration of NaCl (0.25 M to 0.4 M) than that required to elute it from the DEAE-cellulose at pH 7.4.

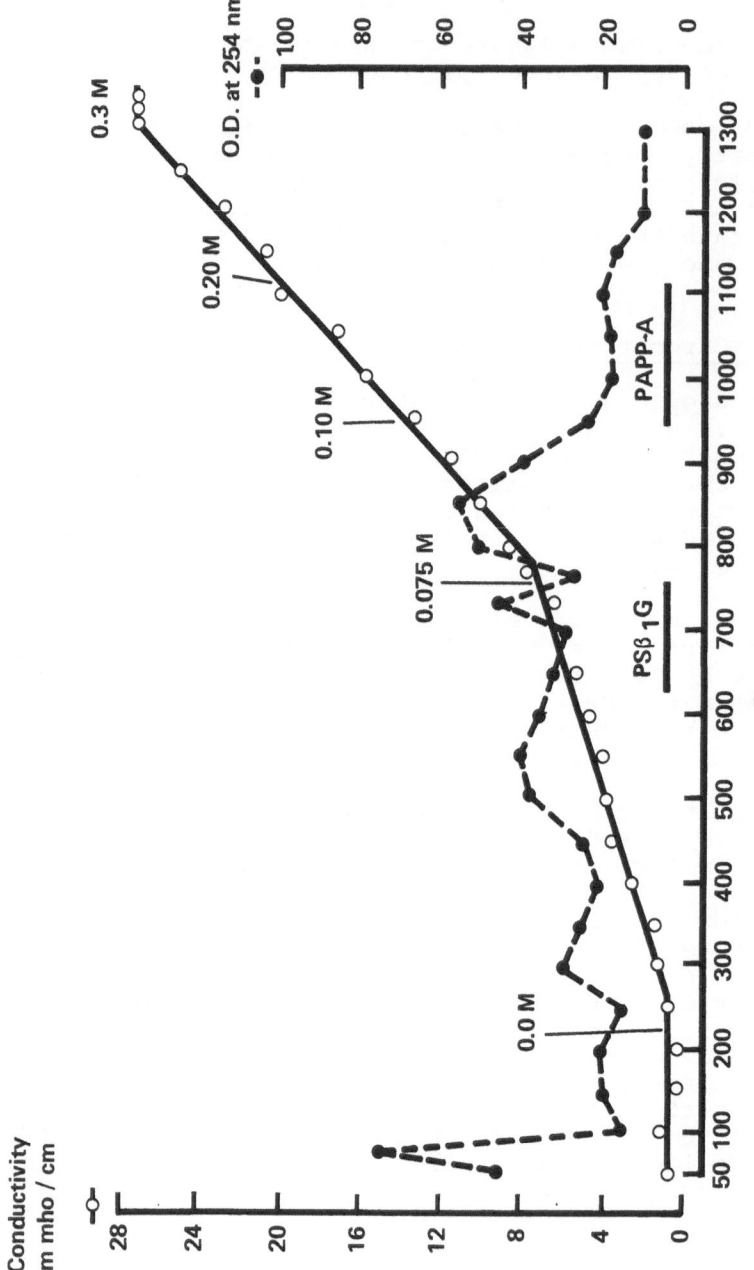

Fig. 7.1. Elution profile of placental proteins from DEAE-cellulose column

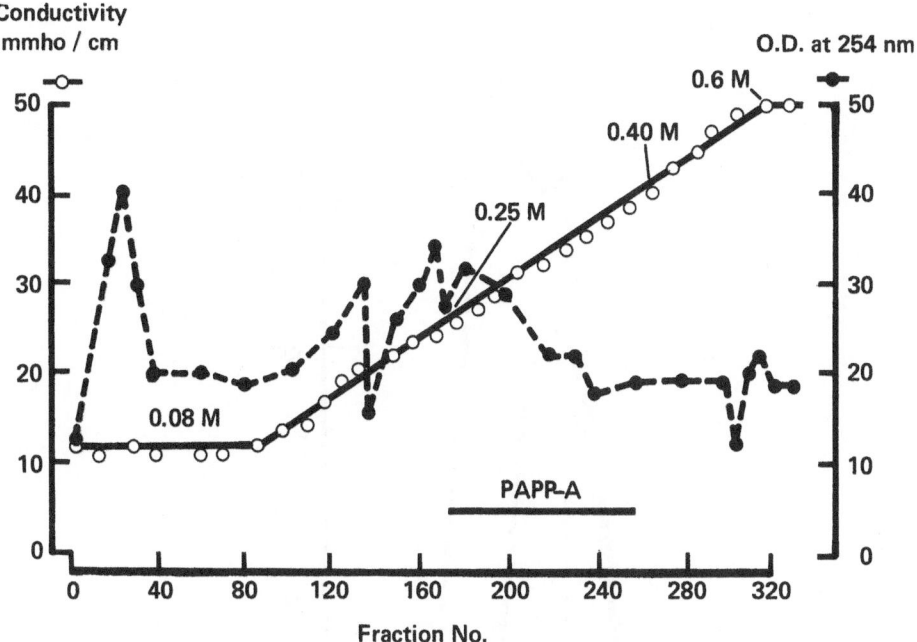

Conductivity mmho / cm

O.D. at 254 nm

0.6 M

0.40 M

0.25 M

0.08 M

PAPP-A

Fraction No.

Fig. 7.2. Elution profile of PAPP-A from a DEAE-Sephadex column

At this stage of the purification, the total amount of protein in the PAPP-A containing fractions represents roughly 0.5% of the proteins present in the starting pregnancy plasma, whereas 36% of the initial amount of PAPP-A was recovered.

7.1.3 Gel Filtration

Having separated the proteins according to charge, we took advantage of knowing the approximate molecular weight of PAPP-A (750,000) in order to select a suitable molecular filtration gel. Sepharose CL 4B was chosen, since it has a "pore size" big enough to allow proteins of molecular weight up to 10^6 to penetrate into the gel. The column used was 2.5x100 cm and was prepared in buffer C (0.01 M sodium phosphate pH 7.0 containing 0.15 NaCl). The volume of samples applied never exceeded 7 ml.

The PAPP-A-containing fractions eluted from DEAE-Sephadex were pooled and dialysed against buffer C, then concentrated by ultrafiltration on a Diaflo membrane (XM 30). The material was applied to the column and eluted with buffer C; the amount of PAPP-A in the fractions was measured by electroimmunodiffusion (*see* Sect. D). The results obtained are shown in Fig. 7.3a.

Three protein peaks were resolved. The first corresponded to the void volume of the column and contained proteins of a molecular weight larger than 10^6 but no PAPP-A. The second peak was resolved as a shoulder on the third peak and contained the maximum amount of PAPP-A whereas the third peak contained less PAPP-A and was composed mainly of contaminating proteins.

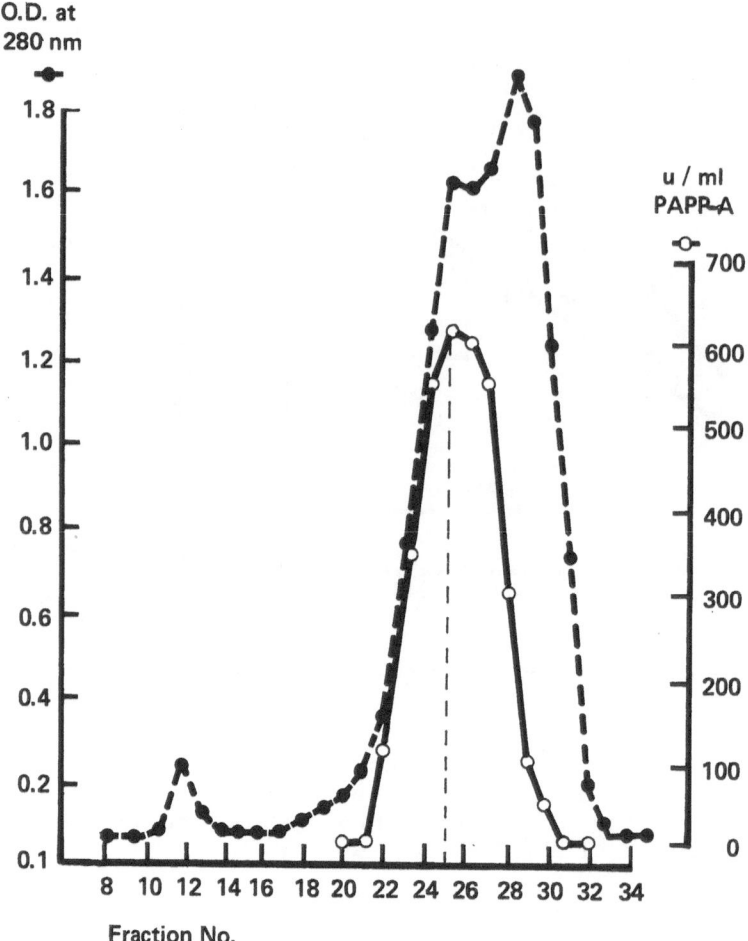

O.D. at
280 nm

u / ml
PAPP-A

Fraction No.

<u>Fig. 7.3.*a*</u> Elution profile of PAPP-A from Sepharose C1 4B

The protein composition of the last two peaks was checked by
submitting aliquots of the peak fractions to 5% polyacrylamide
gel electrophoresis in the presence of SDS (sodium dodecyl
sulphate). This technique separates the proteins according to
molecular size. Five major bands were resolved (Fig. 7.3b), only
the slowest migrating of which corresponded to the distribution
of PAPP-A. Therefore, the PAPP-A preparation still contained
contaminating proteins.

7.1.4 Affinity Chromatography

Because PAPP-A had never been detected in non-pregnancy sera
(LIN et al., 1974b), we attempted to remove unwanted proteins
with an antiserum to normal human serum proteins. The antiserum
was bound to cyanogen-bromide-activated Sepharose as recommended
by the manufacturer (Pharmacia, Uppsala). This material was pre-

Fraction No. 22   23   25   27   29   30

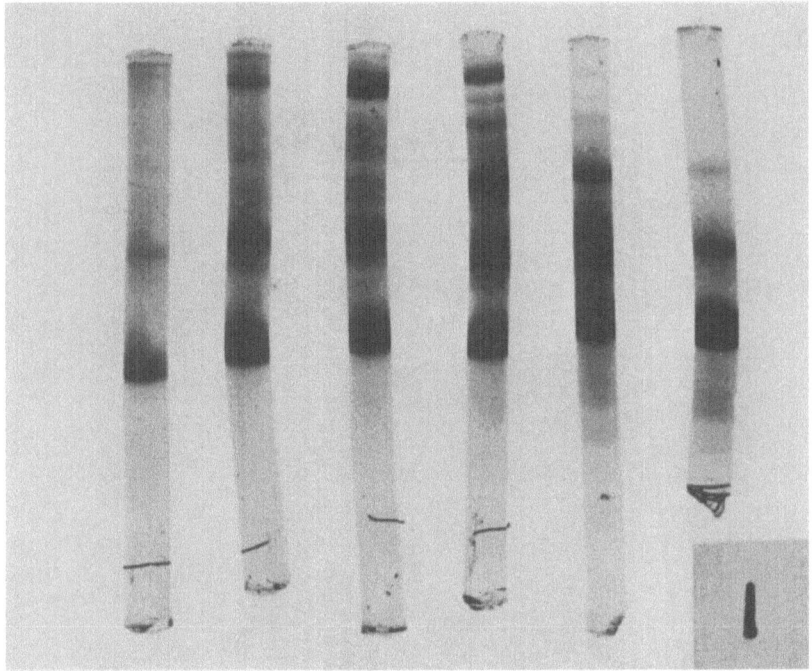

Fig. 7.3.*b* Polyacrylamide gel electrophoresis of PAPP-A peak from DEAE-Sepharose column

pared in buffer D (0.01 *M* sodium phosphate pH 7.0 containing 0.5 *M* NaCl) and used as a 2.5x9 cm column. The PAPP-A containing pool from the gel filtration was dialysed against buffer D and applied to the column; elution was carried out with the same buffer.

A peak of unretarded proteins was eluted in the first 10 fractions and reacted strongly with the PAPP-A antiserum. These fractions were pooled and subjected to repeated runs on the same column to absorb out all contaminating proteins. Between each run the proteins, retained on the column, were washed off with glycine-HCl buffer (pH 2.5). Figure 7.4a shows the elution profile obtained after three successive runs.

The peak of PAPP-A concentration, as measured by electroimmuno-diffusion (see 7.4.2) corresponded to the peak of non-retained proteins, whereas no PAPP-A could be detected in the glycine-HCl wash. When an aliquot of the glycine-HCl wash was submitted to polyacrylamide gel electrophoresis it showed the same major bands as before with the exception of the slow-moving component previously identified as PAPP-A. The PAPP-A fraction showed only a single band (Fig. 7.4b) and therefore appeared to be pure. However, when submitted to immunodiffusion against antiserum to normal human serum a very faint precipitin line could be seen, so that a slight contamination was still present.

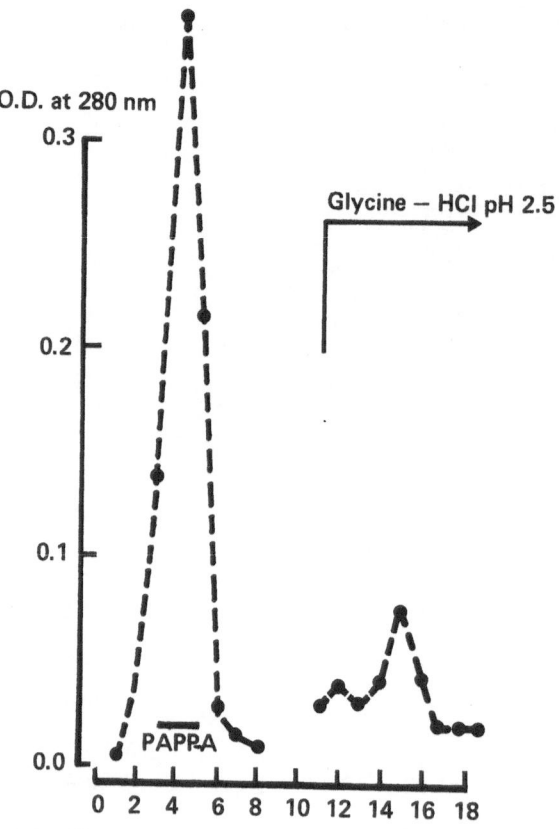

O.D. at 280 nm

Glycine – HCl pH 2.5

PAPP-A

Fraction No.

Fig. 7.4.a Elution profile of PAPP-A from anti-human serum-Sepharose column (negative-affinity column)

7.1.5 Hydroxylapatite Chromatography

Hydroxylapatite is a chromatography medium with interesting adsorption properties. For reasons which are unclear some proteins are capable of binding to this medium whereas others do not.

LIN et al. (1974a) have shown that PAPP-A does not bind to hydroxylapatite; this material was therefore used as a final step in the purification procedure, in the hope that the contaminant might bind.

Hydroxylapatite was equilibrated in buffer E (0.005 M potassium phosphate pH 6.7) and prepared as a column of 2.5x6 cm. The PAPP-A-containing fractions from the preceding affinity column were pooled, dialysed against buffer E and applied to the column. Elution was carried out in two steps. For the first step we used buffer E, which eluted the unretained proteins and for the second step we used an 0.5 M potassium phosphate buffer (pH 6.7) to elute the adsorbed proteins.

Fraction No. 2   3   4   5   6   15

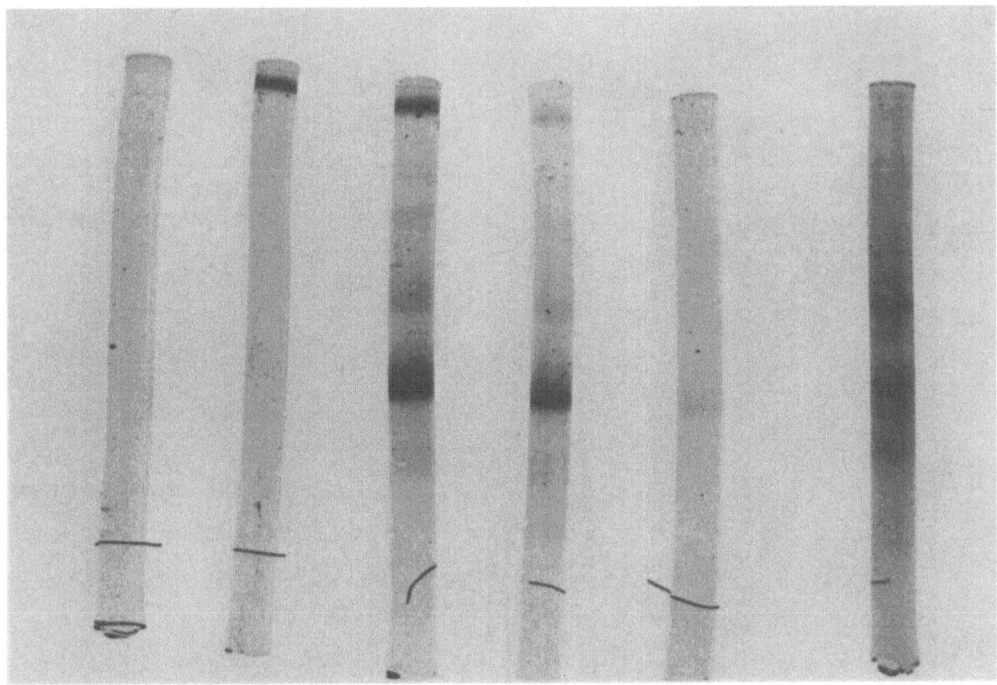

Fig. 7.4.*b* Polyacrylamide gel electrophoresis of PAPP-A peak from negative-
affinity column

As shown on Fig. 7.5.a, one main peak was eluted during the
second step and according to the polyacrylamide gels, (Fig.7.5.b)
PAPP-A and a contaminant were eluted in the same fractions, so
that this purification step did not improve the purity of our
preparation. For unexplained reasons, this result is in contrast
to that of LIN et al. (1974a) who found that PAPP-A did not
bind to hydroxylapatite.

7.1.6 Evaluation of the Purification Method

After each purification step, the concentration of protein,
measured by the method of LOWRY et al. (1951) and the concentration of PAPP-A, measured by electroimmunodiffusion, was determined. The degree of purification and recoveries are shown in
Table 7.1.

PAPP-A had been purified to a considerable degree (294-fold en-
richment) with a yield of around 20%. The hydroxylapatite column
did not improve the purity of our preparation but decreased the
yield to 13%.

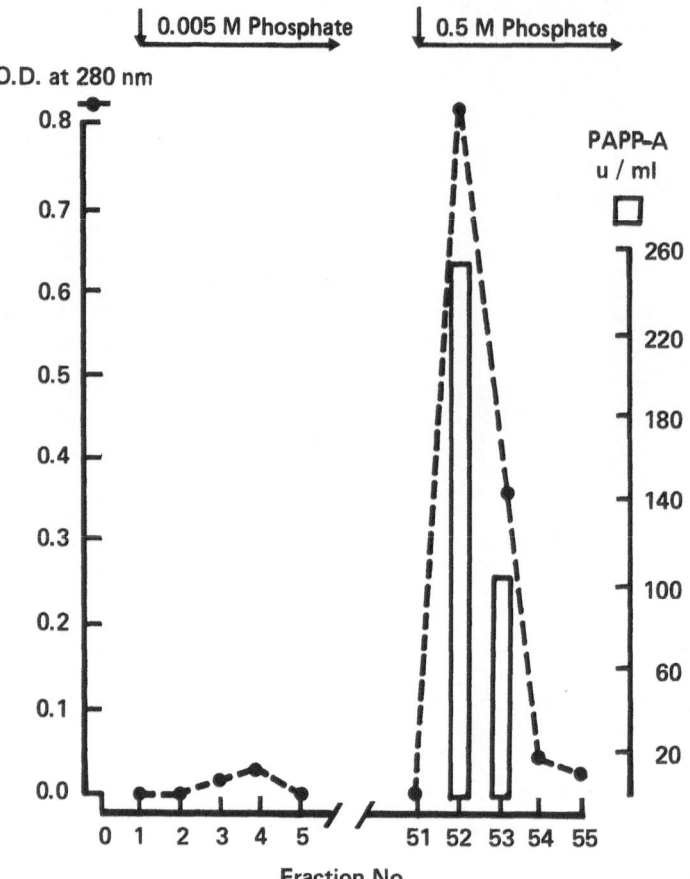

Fig. 7.5.*a* Elution profile of PAPP-A from hydroxylapatite column

7.2 Biochemical Characterization

The electrophoretic mobility of PAPP-A was determined on 1% agarose slabs in barbital buffer pH 8.6. Our preparation migrated as an α_2 globulin, in agreement with the findings of LIN et al. (1974a).

Proteins of known molecular weight were applied to the gel filtration column (Sepharose CL 4B) and eluted in the same manner as described for PAPP-A (*see above*). The elution volumes were calculated and compared to that of PAPP-A (Fig. 7.6).

A molecular weight estimate of 710,000 was obtained for PAPP-A, which is also in good agreement with the value of 750,000 published by LIN et al.(1974a).

Fraction No. 52   53

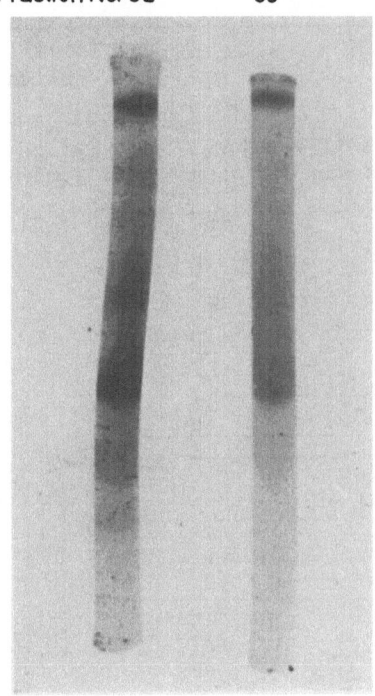

Fig. 7.5.b Polyacrylamide gel electrophoresis
of PAPP-A peak from hydroxylapatite column

7.3 Preparation of Antiserum

An aliquot of the gel filtration pool was extensively dialysed
against a solution of 0.9% NaCl and concentrated to 1 mg/ml by
ultra-filtration.

Rabbits (New Zealand white) were injected subcutaneously at mul-
tiple sites on the back with 120 µg protein homogenised in complete Freund's adjuvant. The rabbits were boosted at monthly
intervals. Blood was collected from a marginal ear vein and the
antiserum thus obtained was absorbed by mixing with an equal
volume of serum obtained from a subject taking oral contraceptives. This serum, rather than normal human serum, was chosen
because oestrogens increase the concentration of some pregnancy-
associated proteins. The mixture of both sera was incubated at
$37^{o}C$ and then at $4^{o}C$ and the precipitate was removed by centrifugation. When this antiserum was tested by immunodiffusion
against normal pregnancy plasma, one precipitin line was resolved which showed complete identity with the precipitin line
obtained with Lin's PAPP-A antiserum.

7.4   Measurement by Immunoelectrophoresis

7.4.1 Standard

A pool of late pregnancy plasma was collected and sodium azide
(0.02% ^{W}/v) and benzamidine hydrochloride 10 mM, (to inhibit

Table 7.1

	Total protein content (mg)	Total PAPP-A content (U)	Purification (-fold)	Recovery (%)
3rd Trimester pregnancy plasma	138380	125400	1	100
60% (v/v) $(NH_4)_2SO_4$ precipitate				
DEAE-cellulose DE32	1959	75900	42.6	60.5
DEAE-Sephadex A50	730	44800	67.4	35.7
Sepharose CL 4B	177	35478	220.0	28.3
Anti-human serum-Sepharose	92	24510	294.1	19.6
OH Apatite	2.0	4218	280.7	12.7
Anti-human serum Sepharose	0.2			

proteolytic enzyme activity, (ENSINCK et al., 1972) were added. This solution was divided in small portions and stored at minus 20^oC. Each millilitre of this standard pregnancy pool was ar- bitrarily designated as containing 100 U PAPP-A. When Lin's standard was compared to ours it was found that Lin's standard read 119 $\pm$ 12 u/ml against 100 for our standard.

7.4.2 Immunoelectrophoresis

This was performed by a modification of the 'rocket' immunoelec- trophoretic technique of LAURELL (1966). Agarose (1%) was dis- solved in barbital buffer (0.05 M pH 8.6) in a boiling water bath. The solution was cooled to 55^oC and antiserum added (final concentration 0.8%^{V}/v). This mixture was made into a slab and left to gel. A row of wells was pierced in the gel and the plate transferred (wells towards the cathode) to an electrophoresis tank filled with the barbital buffer. The standards and the samples were applied in duplicate into the wells and the elec- trophoresis run at 5v/cm for 16 h under cooling. The plate was then washed, dried and stained with Coomassie brilliant blue (R250). The height of the rockets obtained was measured using a photographic enlarger.

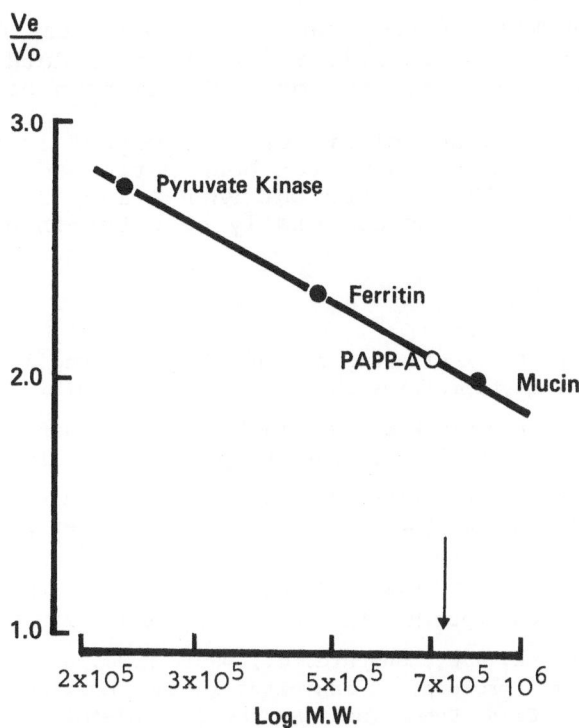

Fig. 7.6. Calibration of the Sepharose Cl 4B column

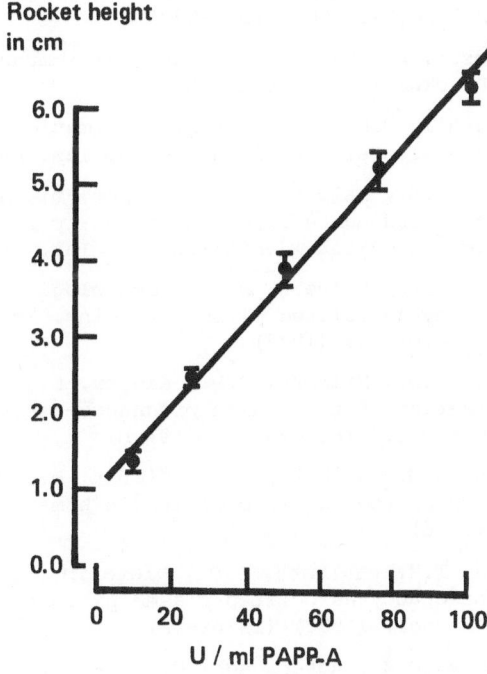

Fig. 7.7. Standard curve for PAPP-A

The mean of the standard curves from 10 different experiments is shown in Fig. 7.7, and from these data a mean interassay coefficient of variation of 4.9% has been calculated.

The reproducibility of this method is thus quite acceptable. No measurements have yet been performed to determine the sensitivity of this method but PAPP-A levels from the 30th week of pregnancy onwards can easily be determined.

References

Bohn, H.: Isolierung und Charakterisierung des schwangerschafts-spezifischen β_1-glykoproteins. Blut <u>24</u>, 292-302 (1972)

Bohn, H., Winckler, W.: Isolierung und Charakterisierung des Plazenta-Proteins PP5. Arch. Gynäkol. <u>223</u>, 179-186 (1977a)

Bohn, H., Winckler, W.: Isolierung und Charakterisierung und quantitative immunologische Bestimmung von zwei neuer Serumproteinen (HPG-1 und HPG-2). Blut <u>35</u>, 305-315 (1977b)

Davies, B.J.: Disc Electrophoresis - II Method and application to human serum proteins. Ann. N.Y. Acad. Sci. <u>121</u>, 404-427 (1964)

Ensinck, J.W., Shepard, C., Dudl, R.J., Williams, R.H.: Use of benzamidine as a proteolytic inhibitor in the radioimmunoassay of glucagon in plasma. J. Clin. Endocrinol. Metab. <u>35</u>, 463-467 (1972)

Gall, S.A., Halbert, S.P.: Antigenic constituents in pregnancy plasma which are undetectable in normal non-pregnant female or male plasma. Int. Arch. Allergy <u>42</u>, 503-515 (1972)

Hudson, L., Hay, F.C.: Practical immunology, 1st ed., pp. 10-124. Oxford: Blackwell Scientific 1976

Laurell, C.B.: Quantitative estimation of proteins by electrophoresis in Agarose Gel containing antibodies. Anal. Biochem. <u>15</u>, 45-52 (1966)

Lin, T.M., Halbert, S.P., Kiefer, D., Spellacy, W.N.: Three pregnancy-associated human plasma proteins; purification, monospecific antisera and immunological identification. Int. Arch. Allergy <u>47</u>, 35-53 (1974a)

Lin, T.M., Halbert, S.P.: Immunological comparison of various human pregnancy associated plasma proteins. Int. Arch. Allergy. Appl. Immunol. <u>48</u>, 101-115 (1975)

Lin, T.M., Halbert, S.P., Kiefer, D., Spellacy, W.N., Gall, S.: Characterization of four human pregnancy-associated plasma proteins. Am. J. Obstet. Gynecol. <u>118</u>, 223-236 (1974b)

Lin, T.M., Halbert, S.P., Kiefer, D.: Quantitative analysis of pregnancy-associated plasma proteins in human placenta. J. Clin. Invest. <u>57</u>, 466-472 (1976)

Lin, T.M., Halbert, S.P., Kiefer,D.: Characterization and purification of pregnancy-associated plasma protein B (PAPP-B). Int. Arch. Allergy Appl. Immunol. (1978) (in press)

Lowry, O.M., Rosebrough, N.J., Farr, A.L., Randall, R.J.: Protein measurement with the folin phenol reagent. J. Biol. Chem. <u>193</u>, 265-275 (1951)

Weber, K., Pringle, J.R., Osborn, M.: Measurement of molecular weights by electrophoresis on SDS - Acrylamide gel. Methods Enzymol. <u>26</u>, 3-27 (1972)

8 Studies on SP$_1$ and PP5 in Early Pregnancy

J.G.Grudzinskas, E.A.Lenton, and B.C.Obiekwe

The placenta is capable of producing a wide variety of protein
molecules which are secreted into the maternal circulation.
These substances have been widely used as diagnostic markers
in early pregnancy and as an index of fetal well-being. General
aspects of the function of these proteins and their clinical
application have been described in the first chapter.

The identification, isolation and purification of a new genera-
tion of placental or trophoblast-'specific' proteins has stimulated much work on the clinical use of the measurement of these
molecules during pregnancy. In this chapter the clinical studies
on pregnancy-specific β_1 glycoprotein (SP$_1$) in early pregnancy
and abnormalities of early pregnancy are reviewed. In addition,
preliminary data are presented on circulating levels of placental protein 5 (PP5) throughout pregnancy.

8.1 Pregnancy-Specific β_1 Glycoprotein

8.1.1 Synthesis and Chemistry of SP$_1$

The isolation and physicochemical properties of SP$_1$ have been
discussed in earlier chapters. A major current problem is nomenclature, since a plethora of names and abbreviations has
been proposed, causing confusion for those unfamiliar with recent advances in this field (Table 8.1).

The placenta at term contains 30 mg SP$_1$. Gel filtration chromatography (Sephadex G 200) has shown a molecular weight of 90-
100,000 daltons for circulating SP$_1$ and for the bulk (90%) of
placental SP$_1$. The remaining 10% of placental SP$_1$ is present in
a larger form of 150,000-200,000 daltons (AL ANI et al., 1978).

Localisation studies using indirect immunofluorescence have demonstrated SP$_1$ in the cytotrophoblast as well as the syncytiotrophoblast at 6-12 weeks gestation (TATARINOV et al., 1976a).
Further studies using transmission electron microscopy, enzyme-
bridge immunoperoxidase techniques, and indirect immunofluorescence have shown that SP$_1$ is predominantly found in the cytoplasm of the syncytiotrophoblast (HORNE et al., 1976; LIN and
HALBERT, 1926; HORNE et al., 1977a). SP$_1$ has also been identified
in the syncytiotrophoblast as early as 2-3 weeks after conception (HORNE et al., 1977a).

8.1.2 Function of SP$_1$

The function of SP$_1$ is unknown. It has been suggested that SP$_1$
might be a regulatory agent in carbohydrate metabolism (TATRA
et al., 1976a), a carrier protein for oestrogen (BOHN and KRANZ,
1973), or an immunosuppressive agent (CERNI et al., 1977).

Table 8.1. Schwangerschaft-spezifisches (pregnancy-specific) β_1 glycoprotein-1 (SP$_1$)

Synonym	Abbreviation	Reference
Pregnancy-specific β_1 globulin	–	TATARINOV and MASYUKEVICH, 1970
Trophoblast β globulin	TBG	TATARINOV et al., 1975
Pregnancy-specific β_1 globulin	B$_1$GP	TATARINOV et al., 1976b
Trophoblast-specific β_1 globulin	TBG	TATARINOV and SOKOLOV, 1977
Trophoblast-specific β glycoprotein	TSG	TATARINOV, 1978
Schwangerschaft spezifische β_1 glycoprotein-1	SP$_1$	BOHN, 1971
Pregnancy-associated plasma protein-C	PAPP-C	LIN et al., 1974
Pregnancy-specific β_1 glycoprotein	PSβG	HORNE et al., 1976
β_1 glycoprotein	β_1SP$_1$	SEARLE et al., 1978

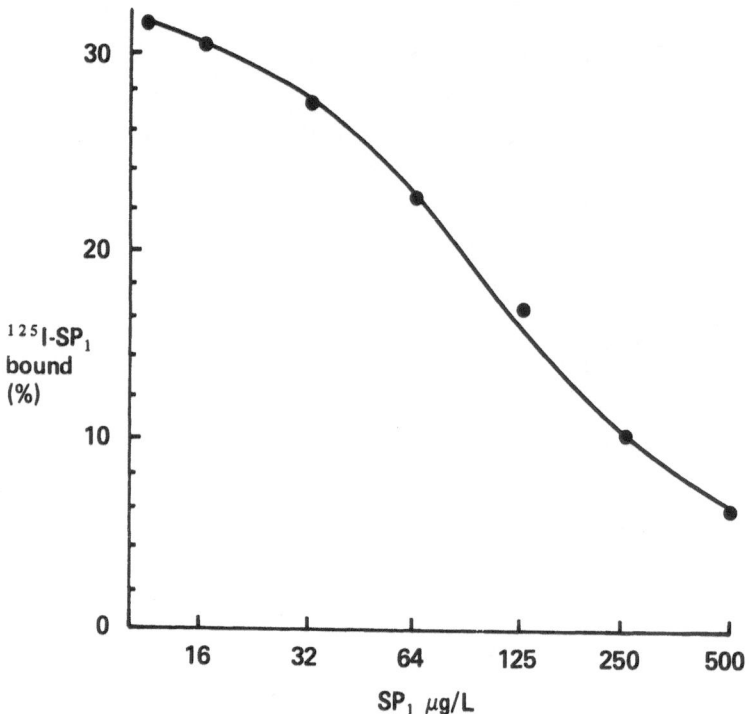

Fig. 8.1. Standard dose-response curve for SP_1 using ^{125}I-SP_1 and rabbit antiserum to SP_1

8.1.3 Measurement of SP_1

As the biological function of SP_1 is unknown the measurement depends on the use of immunological techniques (HORNE et al., 1977b). For studies in early pregnancy when levels are relatively low, sensitive radioimmunoassays have been developed (Fig. 8.1) (GRUDZINSKAS et al., 1977a,b; SORENSEN et al., 1977; SEPPÄLÄ et al., 1978a).

8.1.4 Distribution of SP_1

In common with other placental proteins SP_1 is selectively secreted into the maternal compartment (Fig. 8.2) (TATRA et al., 1976b; SORENSEN et al., 1977; GRUDZINSKAS et al., 1978a). Thus levels in the maternal circulation are several orders of magnitude greater than those in the fetal circulation and amniotic fluid. The concentration of SP_1 excreted in its native form into the urine is approximately 1% of the maternal circulating SP_1 (GRUDZINSKAS et al., 1978a).

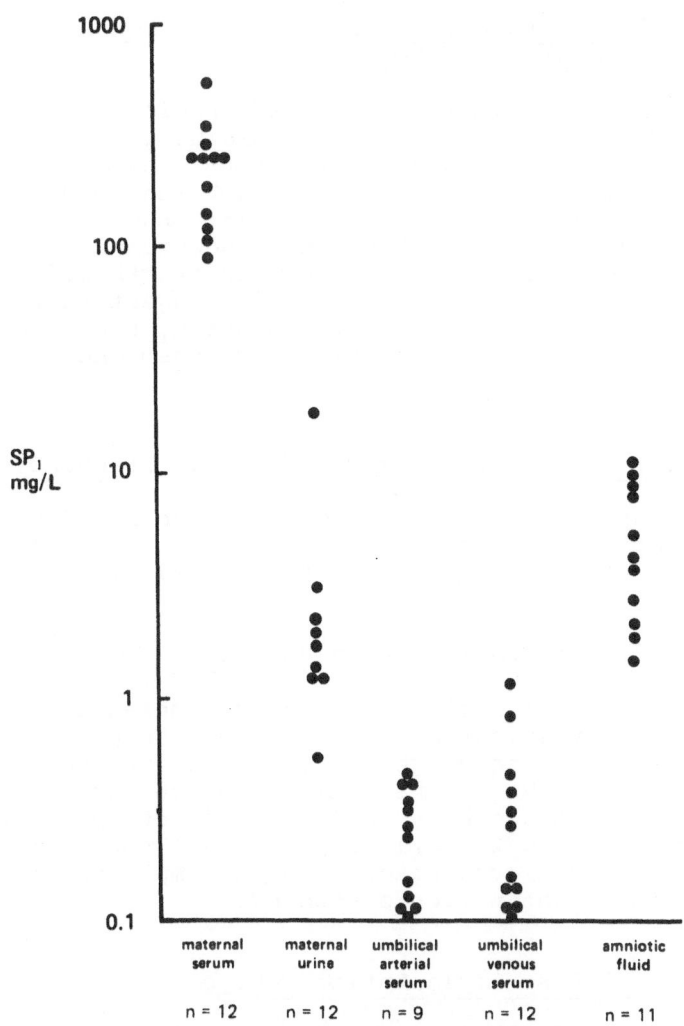

Fig. 8.2. Levels of SP$_1$ in late pregnancy in a variety of fetal and maternal compartments. Reprinted with permission from *Obstetrics and Gynecology* (GRUDZINSKAS et al., 1978)

8.1.5 Circulating Levels of SP$_1$ in Normal Non-Pregnant Subjects

Evidence for the presence of small amounts of SP$_1$ in the normal non-pregnant state is inconclusive. Negative results have been shown using assays with sensitivity limits of 20 µg/litre (GRUDZINSKAS et al., 1977a) and 1 µg/litre (TATARINOV and SOKO-LOV, 1977). However, a recent report (SEARLE et al., 1978) describes the detection of serum levels (3-20 µg/litre) in 12 of 94 normal volunteers. Non-specific binding effects are a well-recognised phenomenon at or near the detection limits of sensitive RIAs and until this possibility is excluded the question of circulating SP$_1$ in non-pregnant subjects must remain controversial.

8.1.6 Circulating Levels of SP$_1$ in Non-Pregnant Subjects with Neoplastic Disease

SP$_1$ has been detected in the circulation of patients with a wide range of non-trophoblastic neoplasms, including carcinoma of the testis, lung, gastrointestinal tract,breast and ovary (JOHNSON et al., 1977; TATARINOV and SOKOLOV, 1977; HORNE et al., 1977b; SEARLE et al., 1978; GRUDZINSKAS et al., 1978b; CROWTHER et al., 1978). This phenomenon is under investigation as a potential marker of tumour growth and response to therapy. However, Searle et al., (1978) suggest that SP$_1$ measurements should be interpreted with caution, since they observed non-specific binding effects in at least one patient with mammary carcinoma.

8.1.7 Circulating Levels of SP$_1$ in Pregnancy

The use of a specific and sensitive RIA for SP$_1$ has permitted detection of circulating levels of this molecule soon after conception (GRUDZINSKAS et al., 1977a).

8.1.7.1 Pre-Implantation Embryos

The production of SP$_1$ by the preimplantation embryo is currently being investigated following extracorporeal fertilisation in women with occlusive fallopian tube pathology (FERGUSON et al., 1978). Assays of the medium from cultures of embryos at the 4-8 cell stage have failed to detect SP$_1$ in concentrations greater than 20 µg/litre. However, continuation of pregnancy was not achieved after surgical implantation of the conceptus in the eight patients studied.

8.1.7.2 Post-Implantation Embryos

In a carefully documented group of subjects plasma levels of SP$_1$ greater than 15 µg/litre were detected 18-23 days after the luteinizing hormone (LH) surge (GRUDZINSKAS et al., 1977b). Thereafter there was a uniform exponential rise in circulating SP$_1$, the doubling time being 2-3 days (Fig. 8.3). The rate of increase declined slightly at 40 days after the LH surge. By 10 weeks gestation the concentration of SP$_1$ is 10-20 µg/litre (LENTON et al., 1978), and in the remaining 30 weeks of pregnancy the levels rise progressively to a plateau of 200-300 µg/litre (GORDON et al.,1977).

The pattern of circulating human chorionic gonadotrophin (hCG) levels, as measured by an assay specific to the β-subunit was similar to that for SP$_1$ up to 40 days (Fig. 8.4). Thereafter the production of hCG decreased, resulting in the characteristic plateau 8-11 weeks after conception. The greater sensitivity of current hCG RIAs permits earlier detection of the molecule (10-16 days after the LH surge). However, the similarity in the pattern of increase and the molar concentrations of the two

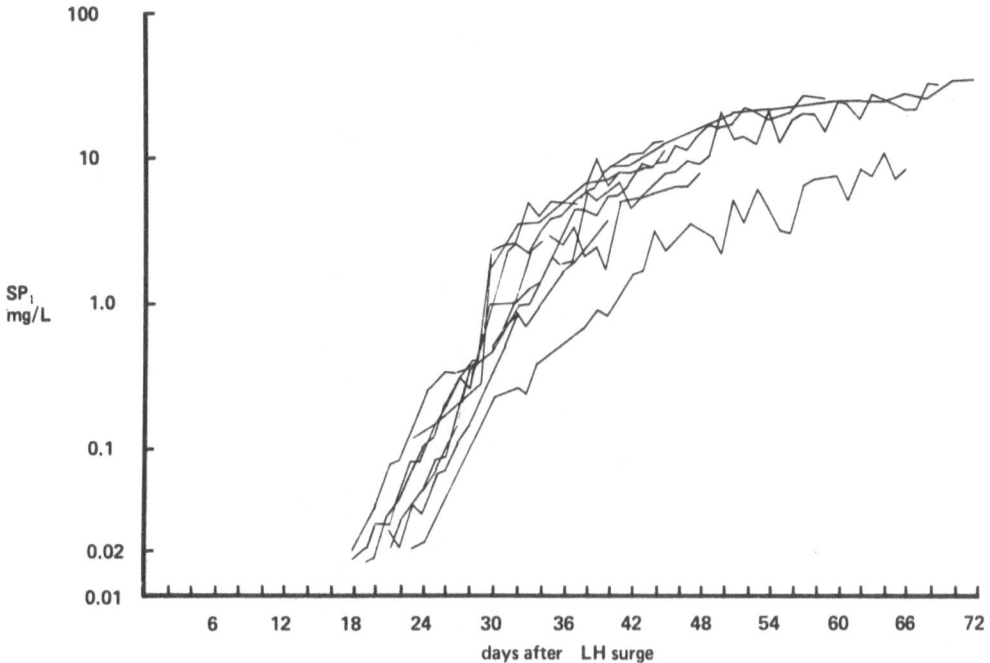

Fig.8.3. Circulating SP_1 in nine normal pregnant subjects

materials suggest that the limitations of the SP_1 assay are technical rather than biological, and that with further refinement of the method the SP_1 assay will be equivalent to that of hCG.

The concentration of SP_1 in maternal urine is less than that in serum and detection of early pregnancy by SP_1 assay in urine is delayed by 14-21 days (GRUDZINSKAS et al., 1977a).

8.1.8   Circulating Levels of SP_1 in Abnormalities of Early Pregnancy

8.1.8.1 Occult Pregnancy

Occult pregnancy is defined as a pregnancy of which neither the woman nor her doctor are aware. The failure to detect pregnancy may be due to either the inability to measure minute amounts of circulating hCG at the time of implantation, or because the patient has been given hCG for ovulation induction or early pregnancy support, thereby rendering hCG determination as a diagnostic test of pregnancy invalid. It has been suggested that as many as 78% of all conceptions may end in abortion (ROBERTS and LOWE, 1975), and that this pregnancy wastage occurs mainly during the occult pregnancy stage (HERTIG, 1975; BLOCH, 1976; BRAUNSTEIN et al., 1977). However, as the diagnosis of occult pregnancy and subsequent subclinical abortion is hindered by the cross-

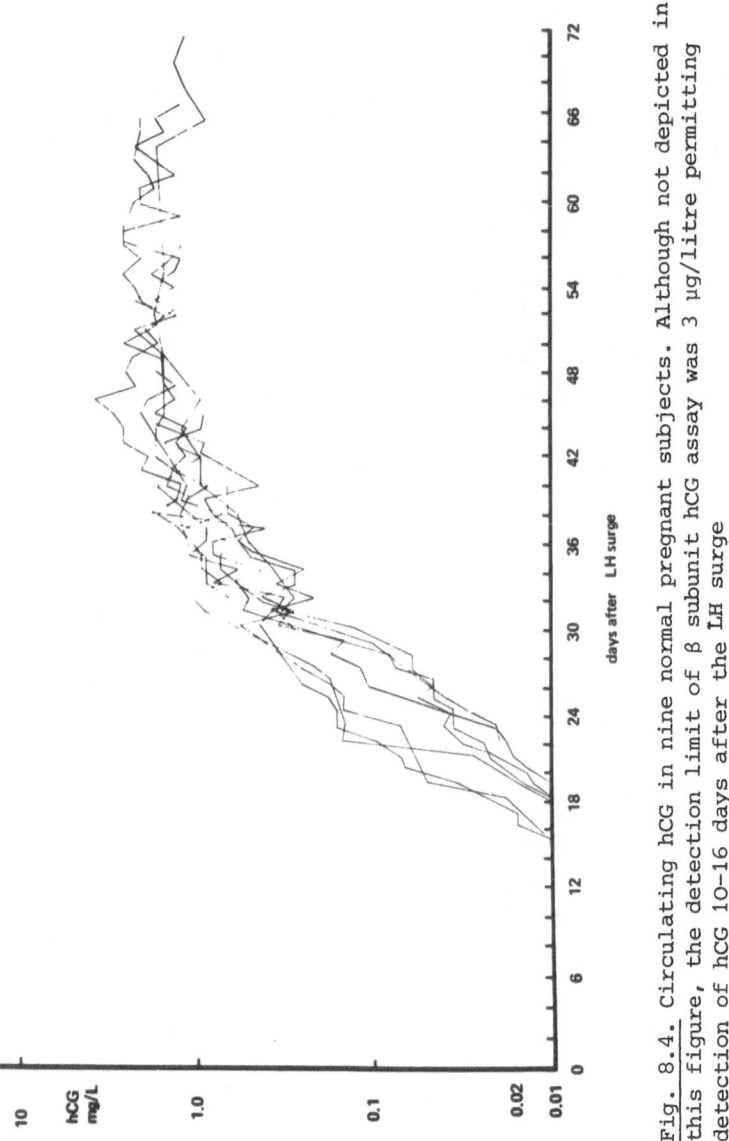

Fig. 8.4. Circulating hCG in nine normal pregnant subjects. Although not depicted in this figure, the detection limit of β subunit hCG assay was 3 μg/litre permitting detection of hCG 10-16 days after the LH surge

reaction of sensitive hCG assays with pituitary gonadotrophins, the hypothesis of Roberts and Lowe has been difficult to evaluate (LANDESMAN and SAXENA, 1976).

SEPPÄLÄ et al. (1978a) have described the detection of both plasma hCG and SP_1 12 days after ovulation in one patient in whom menses was delayed by 7 days. In two other subjects only plasma hCG was detected. These observations suggest that SP_1 measurement might be useful in the diagnosis of occult pregnancy.

8.1.8.2 Anembryonic Pregnancy

Circulating levels of SP_1 were depressed in seven of eight patients with anembryonic pregnancy (blighted ovum) (BENNETT et al., 1978). This observation demonstrates that placental production of SP_1 is independent of the presence of a fetus.

8.1.8.3 Spontaneous Abortion

The measurement of circulating SP_1 in patients with threatened abortion might be useful in predicting the outcome of the pregnancy in a manner similar to that previously described for human placental lactogen (hPL) (NIVEN et al., 1972; GAROFF and SEPPÄLA, 1975). Data from a preliminary study have shown a substantial overlap of plasma SP_1 levels in pregnancies which will continue to term and those which will abort (ETHERIDGE et al., 1978). Whether this is due to the small numbers of subjects examined or to some basic defect in the methology is at present uncertain. However, SCHULTZ-LARSEN and HERTZ (1978) have reported that abortion could be predicted with a high degree of accuracy (89%) if a single low SP_1 value was found, and that the predictive value was higher if serial determinations were obtained.

8.1.8.4 Ectopic Gestation

Circulating levels of SP_1 have been described in association with tubal ectopic gestation (GRUDZINSKAS et al., 1977a).

8.1.8.5 Termination of Pregnancy

MANDELIN et al. (1978) measured serum levels of SP_1 and hCG in 103 women undergoing therapeutic abortion induced by vaginal prostaglandin suppositories. After expulsion of the products of conception, SP_1 and hCG showed a parallel decline in some subjects, while in others, SP_1 seemed to persist for up to 4 weeks. This provides evidence for discordance between SP_1 and hCG levels in early pregnancy.

8.1.8.6 Hydatidiform Mole and Choriocarcinoma

SP_1 has been demonstrated in hydatidiform mole and choriocarcinoma using histochemical techniques (TATARINOV et al., 1976b; HORNE et al., 1977a). In addition SP_1 has been detected in mole tissue homogenate and mole vesicle fluid (GRUDZINSKAS and GORDON, unpublished observations). Circulating SP_1 has been described in the majority of patients with hydatidiform mole and chorio-carcinoma (TATARINOV et al., 1974, 1975, 1976b, SEPPÄLÄ et al., 1978b; SEARLE et al., 1978). Recent reports have described the presence of circulating SP_1 in the absence of hCG in some patients with choriocarcinoma (SEPPÄLÄ et al., 1978b; SEARLE et al., 1978). This observation further emphasises the possible discordance between SP_1 and hCG levels and suggests that a significant improvement in the detection of subclinical tumour growth may be possible if both SP_1 and hCG are measured.

8.2 Placental Protein 5

Placental protein 5 (PP5) is one of a growing list of proteins isolated from placental extracts by BOHN (1976). It is considered to be a tissue protein which is a constituent of placental tissue not secreted in significant amounts into the maternal circulation.

8.2.1 Chemistry and Synthesis of PP5

PP5 is a β glycoprotein with a carbohydrate content of 19% and a molecular weight of 36,600 daltons. The content in the term placenta is 1-2 mg, and it is localised principally in the syncytiotrophoblast (JONES and CHAPMAN, 1978; HEYDERMAN et al., 1978). In earlier studies using gel diffusion PP5 could not be detected in the maternal circulation in concentrations greater than 100 μg/litre (BOHN, 1976).

8.2.2 Measurement of PP5

The development of a sensitive and specific RIA for PP5 has permitted detection of circulating levels of PP5 throughout pregnancy (OBIEKWE et al., 1978). Purified PP5 antigen and anti-PP5 antiserum were provided by Dr. Hans Bohn, Behringwerke, Marburg. The dynamic range of the assay is 2-500 μg/litre (Fig. 8.5). No cross-reaction was demonstrated with α-feto-protein, SP_1, hCG, and pituitary gonadotrophins. However, cross-reaction was observed with hPL (0.001%); this possibly indicates trace contamination of purified hPL with PP5, since I^{125}-labelled PP5 failed to bind to a monospecific antiserum raised against hPL.

8.2.3 Circulating Levels of PP5 in Non-Pregnant Subjects

Circulating levels of PP5 greater than 1 μg/litre were not detected in 50 normal non-pregnant subjects.

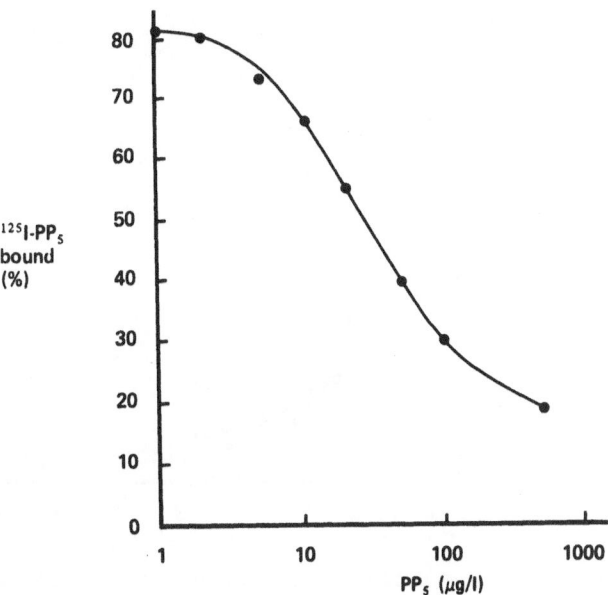

Fig. 8.5. Standard dose-response curve for PP5 which was determined by means of ^{125}I-PP5 and rabbit antiserum to PP5

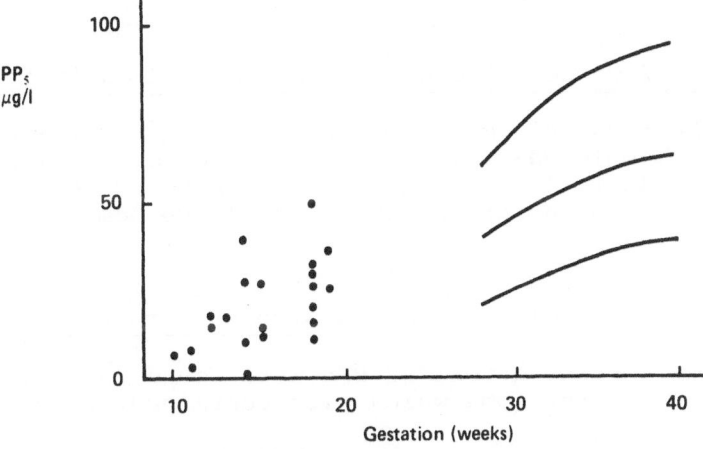

Fig. 8.6. Circulating levels of PP5 in normal pregnancy (·). PP5 concentration for individual subjects, The lines present the 10th, 50th and 90th centiles of 560 PP5 estimation obtained from 280 patients

8.2.4 Circulating Levels of PP5 in Pregnancy

PP5 was detected in the serum of all pregnant subjects from 8 weeks gestation onwards (Fig. 8.6). PP5 levels rise throughout pregnancy and plateau at about 36 weeks, the median level near term being 45 µg/litre. Trace amounts of PP5 have been detected in maternal urine and amniotic fluid. Circulating levels of PP5

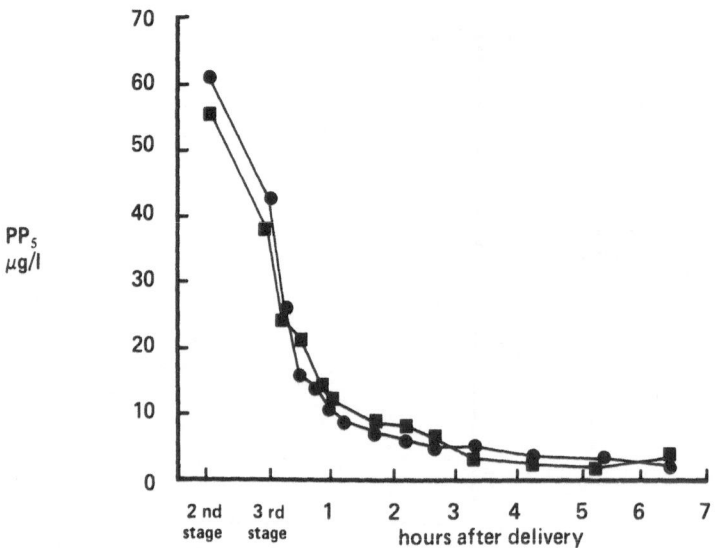

Fig. 8.7. Circulating levels of PP5 after delivery in two patients

have also been detected in cases of ectopic pregnancy and threatened abortion. The half-life of PP5 following delivery is
15-30 min (Fig. 8.7).

8.2.5 Circulating Levels of PP5 in Hydatidiform Mole

The range of serum levels of PP5 in eight subjects with hydatidiform mole was 2-46 µg/litre. After evacuation of the mole, PP5
levels fell more rapidly than those of hCG. PP5 is present in
high concentrations in molar tissue homogenates and mole vesicle
fluid.

8.2.6 Circulating Levels of PP5 in Testicular Teratoma

Low concentrations of PP5 were found in a homogenate of a testicular teratoma which also contained SP_1 and hCG.

8.2.7 Comparison Between Levels of PP5 in Plasma and Serum Samples

A consistent difference was noted in the PP5 levels in plasma
and serum samples from the same subject. The serum levels were
1.5-3-fold greater than the levels present in plasma collected
in lithium-heparin (12.5 U/ml), whereas the presence of oxalate
or citrate in the plasma produced intermediate PP5 values. Identical PP5 levels were obtained in glass and plastic tubes.

Investigations of the heparin effect showed that the addition
of heparin to serum samples caused a marked reduction in PP5

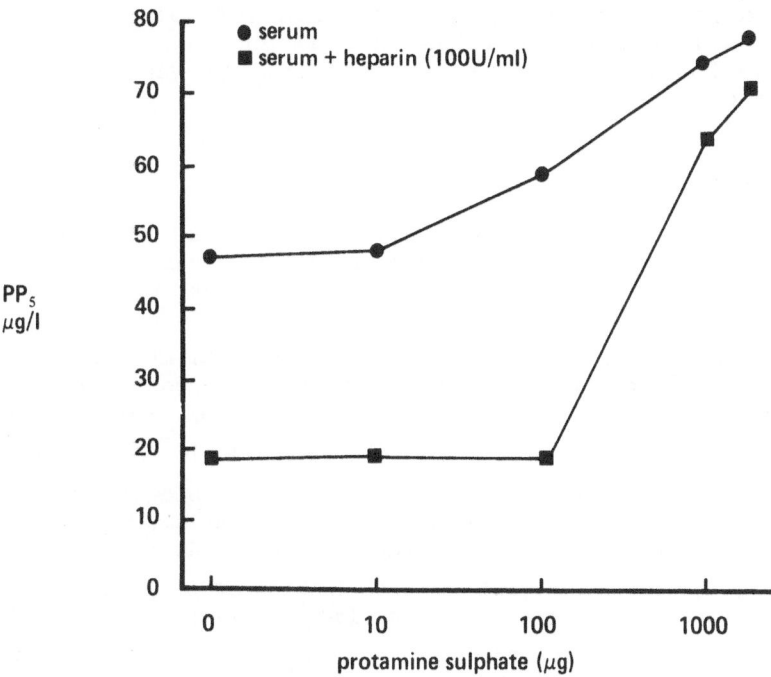

Fig.8.8. The effect of heparin and protamine sulphate on circulating PP5

values. The effects of heparin could be reversed by the addition of protamine sulphate (Fig. 8.8). Thus heparin caused an apparent enhancement of antibody binding and falsely lowered PP5 levels whereas protamine had the opposite effect. The effect was not observed when PP5 standards were assayed in the absence of pregnancy sera, indicating a difference between purified and circulating PP5.

8.3   Conclusions

The development of a specific and sensitive RIA for SP_1 has permitted the detection of small amounts of circulating SP_1 in very early pregnancy, in a variety of maternal and fetal compartments, in abnormalities of early pregnancy and in malignant diseases.

Clinical studies indicate that the measurement of circulating SP_1 may be a useful adjunct in the diagnosis of pregnancy and its abnormalities. This measurement may be the only biochemical method of pregnancy diagnosis if hCG has been given to the patient. The discordance of SP_1 and hCG levels in trophoblastic disease demonstrates the value of SP_1 measurement in this condition in the detection of subclinical disease or the monitoring of therapy.

Finally, the development of a RIA for PP5 has permitted the study of circulating PP5. Low circulating concentrations of PP5 suggest that this molecule is a storage protein which incidentally appears in the peripheral circulation, in contrast to the high blood levels of SP_1. The initial encouraging results are a stimulus for further evaluation of this potential new marker substance.

References

Al-Ani, A.T.M., Grudzinskas, J.G., Gordon, Y.B., Chard,T.: Molecular forms of pregnancy specific β_1 glycoprotein in a variety of maternal and fetal compartments. Vol. II. Proc. 6th meeting int. res. group carcino-embryonic proteins, 1978

Bennett, M.J., Grudzinskas, J.G., Gordon, Y.B., Turnbull, A.C.: Circulating levels of alpha fetoprotein and pregnancy specific β_1 glycoprotein in pregnancies without an embryo. Br. J. Obstet. Gynaecol. 85, 348-350 (1978)

Bloch, S.K.: Occult pregnancy. Obstet. Gynecol. 48, 365-368 (1976)

Bohn, H.: Nachweis und Charakterisierung von Schwangerschaft-Proteinen in der menschlichen Plazenta, sowie ihre quantitative immunologische Bestimmung im Serum schwangerer Frauen. Arch. Gynäkol. 210, 440-457 (1971)

Bohn, H.: Isolation and characterization of placental specific proteins SP_1 and PP5. Protides Biol. Fluids 24, 117-124 (1976)

Bohn, H., Kranz, T.: Identification of pregnancy associated β_1 glycoprotein with the steroid binding globulin. Arch. Gynaekol. 215, 63-71 (1973)

Braunstein, G.D., Karow, W.G., Gentry, W.D., Wade, H.E.: Subclinical spontaneous abortion. Obstet. Gynecol. 50, 415-445 (1971)

Cerni, C., Tatra, G., Bohn, H.: Immunosuppression by human placental lactogen and the pregnancy specific β_1 glycoprotein. Arch. Gynaekol. 223, 1-7 (1977)

Crowther, M., Grudzinskas, J.G., Poulton, T., Gordon, Y.B.: The measurement of circulating levels of pregnancy specific β_1 glycoprotein, human chorionic gonadotrophin and human placental lactogen in carcinoma of the ovary. Obstet. Gynecol. (1978) (in press)

Etheridge, R., Grudzinskas, J.G., Gordon, Y.B., Chard, T.: Pregnancy specific β_1 glycoprotein in threatened miscarriage. (unpublished observations) (1978)

Ferguson, I.L.C., Grudzinskas, J.G., Brush, M.G., Gordon, Y.B.: (unpublished observations) (1978)

Garoff, L., Seppälä, M.: Prediction of fetal outcome in threatened abortion by maternal serum lactogen and alphafetoprotein. Am. J. Obstet. Gynecol. 121, 257-261 (1975)

Gordon, Y.B., Grudzinskas, J.G., Jeffrey, D., Chard, T., Letchworth, A.T.: Concentrations of pregnancy specific β_1 glycoprotein in normal pregnancy and in intrauterine growth retardation. Lancet i, 331-333 (1977)

Grudzinskas, J.G., Gordon, Y.B., Jeffrey, D., Chard, T.: Specific and sensitive determination of pregnancy specific β_1 glycoprotein by radioimmunoassay. Lancet i, 333-335 (1977a)

Grudzinskas, J.G., Lenton, E.A., Gordon, Y.B., Kelso, I.M., Jeffrey, D., Sobowale, O., Chard, T.: Circulating levels of pregnancy specific β_1 glycoprotein in early pregnancy. Brit. J. Obstet. Gynaecol. 84, 740-742 (1977b)

Grudzinskas, J.G., Evans, D.G., Gordon, Y.B., Jeffrey, D., Chard, T.: Pregnancy specific β_1 glycoprotein in fetal and maternal compartments. Obstet. Gynecol. 52, 43-46 (1978a)

Grudzinskas, J.G., Coombes, R.C., Ratcliffe, J.G., Gordon, Y.B., Munro Neville, A., Chard, T.: Circulating levels of pregnancy specific β_1 glycoprotein in patients with testicular, bronchogenic and breast carcinomas: (submitted to Lancet) (1978b)

Hertig, A.I.: Implantation of the human ovum. The histogenesis of some aspects of spontaneous abortion. In: Progress in infertility. BEHRMAN, S.J., KISTNER, A.W. (eds.), pp. 411-438. Boston: Little, Brown 1975

Heyderman, E., Grudzinskas, J.G., Bohn, H., Chard, T.:(unpublished observations) (1978)

Horne, C.H.W., Towler, C.M., Pugh-Humphreys, R.G.P., Thomson, A.W., Bohn, H.: Pregnancy specific β_1 glycoprotein - a product of the syncytiotrophoblast. Experientia 32, 1197-1199 (1976)

Horne, C.H.W., Towler, C.M., Milne, G.D.: Detection of pregnancy specific β_1 glycoprotein in formalin fixed tissues. J. Clin. Pathol. 30, 19-23 (1977a)

Horne, C.H.W., Towler, C.M., Jandial, V.: Pregnancy specific β_1 glycoprotein in early pregnancy, Lancet i, 707-709 (1977b)

Johnson, S.A.N., Grudzinskas, J.G., Gordon, Y.B., Al-Ani, A.T.M.: Pregnancy specific β_1 glycoprotein in plasma and tissue extracts in malignant teratoma of the testis. Br. Med. J. i, 951-952 (1977)

Jones, W.R., Chapman, M.: (unpublished observations) (1978)

Landesman, R., Saxena, B.B.: Results of the first 1000 radio receptor assays for the determination of human chorionic gonadotrophin: a new rapid reliable and sensitive pregnancy test. Fertil. Steril. 27, 357-368 (1976)

Lenton, E.A., Grudzinskas, J.G., Sobowale, O., Gordon, Y.B.: (unpublished observations) (1978)

Lin, T.M., Halbert, S.P., Kiefer, D., Spellacy, W.N., Gall, S.: Characterisation of four human pregnancy associated plasma proteins. Am. J. Obstet. Gynecol. 118, 213-236 (1974)

Lin, T.S., Halbert, S.P.: Placental localization of human pregnancy associated plasma proteins. Science 193, 1249-1252 (1976)

Mandelin, M., Rutanen, E.M., Heikinheimo, M., Jalanko, H., Seppälä, M.: Pregnancy specific β_1 glycoprotein and chorionic gonadotrophin levels after first trimester abortions induced by 15-S 15 methyl prostaglandin F vaginal suppositories. Obstet. Gynecol. (1978) (in press)

Niven, P.A.R., Landon, J., Chard, T.: Placental lactogen levels as a guide to outcome of threatened abortion. Br. Med. J. iii, 799-801 (1972)

Obiekwe, B.C., Gordon, Y.B., Grudzinskas, J.G., Pendlebury, D., Bohn, H., Chard, T.: Circulating levels of placental protein 5 (PP5) during pregnancy. Proc. 6th meeting of international research group for carcino-embryonic proteins, Vol. II (1978) (in press)

Roberts, C.J., Lowe, C.R.: 'Where have all the conceptions gone?' Lancet i, 489-499 (1975)

Schultz-Larsen, P., Hertz, J.B.: The predictive value of pregnancy specific
β_1 glycoprotein (SP$_1$) in threatened abortion. Eur. J. Obstet. Gynecol.
Reprod. Biol. (1978) (in press)

Searle, F., Leake, B.A., Bagshawe, K.D., Dent, J.: Serum SP$_1$ pregnancy specific β_1 glycoprotein in choriocarcinoma and other neoplastic disease.
Lancet i, 579-581 (1978)

Seppälä, M., Lehtovirta, P.,Rutanen, E.M.: Detection of chorionic gonadotrophin-like activity in infertile cycles with a short luteal phase. Acta
Endocrinol. (Kbh.) 88, 164-168 (1978a)

Seppälä, M., Rutanen, E.M., Heikinheimo, M., Jalanko, H., Engvall.E.: Detection of trophoblastic tumour activity by pregnancy specific β_1 glycoprotein. Int. J. Cancer 21, 265-267 (1978b)

Sorensen, S., Borggaard, B., Rolff, L.: A radioimmunoassay of the pregnancy
specific β_1 glycoprotein (SP$_1$). Scand. J. Clin. Lab. Invest. 37, 537-543
(1977)

Tatarinov, Y.S.: A new placental protein test for the presence and identification of trophoblastic tumours. Antibiot. Chemother. 22, 125-131 (1978)

Tatarinov, Y.S., Masyukevich, V.N.: Immuno-chemical identification of new
beta$_1$ globulin in the blood serum of pregnant women. Byull. Eksp. Biol.
Med. 69, 66-68 (1970)

Tatarinov, Y.S., Sokolov, A.V.: Development of a radioimmunoassay for pregnancy specific beta$_1$ globulin and its measurement in serum of patients with
trophoblastic and non-trophoblastic tumours. Int. J. Cancer 19, 161-166
(1977)

Tatarinov, Y.S., Mesnyankina, N.V., Nikoulina, D.M., Novikova, L.A., Toloknov,
B.O., Falaleeva, D.M.: Identification immunochimique de la béta globulin
de la 'zone de grossesse' dans le serum de malades attendes de tumeurs
trophoblastiques. Int. J. Cancer 14, 548-554 (1974)

Tatarinov, Y.S., Falaleeva, S., Elgort, D.A., Novikova, L.A., Toloknov, B.O.:
Immunoautoradiographic determination of beta$_1$ globulin in the blood stream
of patients with trophoblastic tumours. Byull. Eksp. Biol. Med. 74, 86-89
(1975)

Tatarinov, Y.S., Falaleeva, D.M., Kalashnikov, V.V., Toloknov, B.O.: Immunofluorescent localization of human pregnancy specific β globulin in placenta
and chorioepithelioma. Nature 260, 263 (1976a)

Tatarinov, Y.S., Falaleeva, D.M., Kalashnikov, V.V.: Human pregnancy specific
beta$_1$ globulin and its relation to chorioepithelioma. Int. J. Cancer 17,
626-632 (1976b)

Tatra, G., Tempfer, H., Placheta, P.: Influence of blood glucose levels on
serum concentrations of pregnancy specific protein SP$_1$ and hPL during the
last trimester of pregnancy. Eur. J. Obstet. Gynecol. Reprod. Biol. 6/2
53-57 (1976a)

Tatra, G., Polak, S., Placheta, P.: Concentration of pregnancy specific B
protein SP$_1$ in amniotic fluid in normal and pathologic pregnancies. Arch.
Gynaekol. 221, 161-166 (1976b)

9 Clinical Aspects of Pregnancy-Specific β_1 Glycoprotein (PSβG)

G. Tatra

Soon after the fundamental studies of BOHN (1971, 1974), the possibility that measurement of pregnancy-specific β_1 glycoprotein (PSβG) could be of importance in the evaluation of normal and complicated pregnancies was explored by several clinical groups. This interest has centred on both early and late pregnancy, and also on production of PSβG by tumours.

9.1 Trophoblastic Tumours

Pregnancy-specific-β_1 glycoprotein has been detected in the trophoblast by immunofluorescent (SEDLACEK et al., 1976, LIN and HALBERT, 1976) and immunoperoxidase (HORNE et al., 1976c) techniques. Ultrastructural studies of first trimester placentae by HORNE et al., (1976c) confirmed that synthesis of PSβG is localized within the syncytiotrophoblast. However, TATARINOV et al. (1976) suggested that production may begin in the Langhans cells and continue in the syncytiotrophoblast, in both normal chorion and trophoblast tumours.

Measurement of hCG in plasma or urine is widely used in the diagnosis and management of trophoblastic tumours. As the concentration of hCG appears to be related to the total mass of tumour cells, the rate of growth or regression of the tumour cell population can be predicted (BAGSHAWE, 1969). Although there is no obvious clinical need to seek for alternative proteins as tumour markers in this situation, it is nevertheless worthwhile to compare the levels of PSβG in the sera of patients with trophoblastic tumours, since such measurements could be of value if discordances were found to have prognostic significance. Recent investigations by SEARLE et al. (1978) showed PSβG concentrations of up to 50 mg/l in the sera of patients with hydatidiform mole, invasive moles, choriocarcinoma or malignant teratoma. In general, hCG and PSβG appeared to move in parallel, but some cases were noted in which the concentrations were not related. Persistence of PSβG after hCG has become undetectable may therefore be a useful marker (SEARLE et al., 1978).

9.2 Non-Trophoblastic Tumours

Ectopic production of so-called trophoblast-specific proteins by non-trophoblastic tumours is now a well recognized phenomenon. Trace amounts of "pregnancy-specific" protein hormones have been demonstrated by radioimmunoassay in the plasma of some patients with various malignant tumours (WEINTRAUB and ROSEN, 1971, BRAUN-STEIN et al., 1973). Using an enzyme-bridge immunoperoxidase technique, HORNE and co-workers demonstrated that 76% of patients with breast cancer had detectable pregnancy-specific-β_1 glyco-protein in the cytoplasm of the tumour cells (HORNE et al., 1976b). It was suggested by HORNE and co-workers that the absence of PSβG had prognostic significance, as women with β_1 glycopro-tein negative cancers apparently had survival times significantly longer than those who were positive (HORNE et al., 1976a). This important finding needs to be confirmed on a larger scale.

Since PSβG is said to have immunosuppressive properties, at least in vitro (JOHANNSEN et al., 1976; CERNI et al., 1977), it seems likely that the production of this protein by malignant tumours might be a means by which the tumour escapes immunologi-cal recognition and continues to grow. If this were the case, the ability of malignant tumours to synthesize proteins with im-munosuppressive properties could ultimately determine the length of survival of the host.

PSβG has been detected in the sera of some patients with malig-nant tumours of the lung, gastrointestinal tract, breast and ova-ry (TATARINOV and SOKOLOV, 1977).

SEARLE et al. (1978) suggested that PSβG concentrations might be elevated in patients with carcinoma of the breast, but such in-creases need to be distinguished from the increased values found in 13% of normal subjects. Though SEARLE's results confirm the report of TATARINOV and SOKOLOV (1977) of increased values in some patients with colorectal cancer, their evidence suggests that this does not correlate with the extent of the disease.

SEARLE et al. (1978) suggest that serum PSβG concentrations are unlikely to be of great value in the detection and monitoring of carcinoma of the breast, large bowel or ovary.

9.3 Antifertility Effect of Immunization to PSβG

It is known that immunization against the β-subunit of hCG or hPL prevents implantation of the embryo in both rodents and rabbits. However, because of the close relationship between hCG and luteinizing hormone, and hPL and growth hormone, cross-reaction with normal products is a potential problem. Since there is no such problem with PSβG, at least on theoretical grounds, immunization against this protein might prove an effective method of fertility control. BOHN and WEINMANN (1974, 1976) have demonstrated in cynomolgus monkeys that passive and active immunization against PSβG leads to abortion. It remains to be seen, therefore, whether immunization against trophoblast-specific products, especially PSβG, is effective in preventing pregnancies in humans.

9.4 Early Pregnancy

The measurement of proteins specifically produced by the placental trophoblast and secreted into the plasma and excreted in the urine of pregnant women is widely used to detect early pregnancy. Since the development of specific and sensitive radioimmunoassay for PSβG (GRUDZINSKAS et al., 1977a; TATARINOV and SOKOLOV, 1977; SØRENSEN et al., 1977), this material has been investigated as such a marker.

PSβG has been detected in plasma 18-23 days after induced or spontaneous ovulation with a sharp increase following the initial appearance (GRUDZINSKAS et al., 1977b). This suggests that the measurement of PSβG may provide valuable additional biochemical evidence of pregnancy, and may also serve to identify abnormalities of early pregnancy such as missed abortion or anembryonic pregnancy which are associated with low levels.

9.5 Second Half of Pregnancy

Our own studies on serum PSβG concentration during the second half of pregnancy have been carried out using single radial immunodiffusion (Partigen plates, Behringwerke). The range of this technique is between 1 and 25 mg/100 ml. The procedure is well suited to routine clinical work, since it is easily and quickly performed and appears to give reproducible results. Values can be read after 24 h.

Using this technique the half-life of PSβG in blood was found to be 34 h. A close positive correlation was found between serum PSβG concentration, placental weight and fetal weight (TATRA et al., 1975).

Serum PSβG concentrations increase steadily after the 20th week of gestation to reach a plateau at the 37th week. The mean value increases from 3.35 mg/100 ml at the 20th week to 15.03 at the 40th week (TATRA et al., 1974).

These values are 30% lower than those reported by TOWLER et al. (1976), which may be due to the use of different preparations of standard PSβG.

Since the physiological role of PSβG during pregnancy is unknown and the factors underlying its production, release and elimination are uncertain, an attempt was made to define mechanisms which might control the secretion or metabolism of PSβG in the last trimester of gestation. Serum levels of PSβG were significantly increased 24 and 48 h after insulin injection. Similarly, glucose loading significantly increased PSβG levels at 24 h, when compared with the concentrations at 60 or 120 min. These results suggest that serum PSβG levels may be related to the blood sugar level and carbohydrate metabolism (TATRA et al., 1976a). However, mean serum PSβG levels did not show any significant change after continuous infusion of a mixture of amino-acids for 4 h (TATRA et al., 1977).

Further studies were carried out to determine if there was any relation between PSβG levels and corticosteroid administration. Women with threatened premature labour received 12 mg 16-β-methylprednisolone orally for 6-10 days; no other medication was given. Serum PSβG levels were not found to be significantly changed (TATRA et al., 1976b). Serum concentrations of PSβG were investigated for circadian variation: the levels were significantly decreased at 24.00 and 04.00 h. the decrease during the night may be explained by transfers between maternal compartments during rest. Serum PSβG concentrations did not vary significantly during labour (TATRA et al., 1976d).

The concentration of PSβG in amniotic fluid was evaluated in normal and pathological pregnancies. The level in normal fluid averaged 0.126 mg/100 ml   (standard error of mean value $\pm$ 0.012) or approximately 1% of the maternal serum concentration. Levels were elevated in cases of diabetes, intrauterine death and Rh-incompatibility. The increase was 42% in diabetes and 63% in Rh-incompatibility; with intrauterine death the PSβG concentration was 12-fold the normal level, probably due to lysis of the membranes. Amniotic fluid PSβG levels may, therefore, be of clinical significance. A concentration of 0.4 mg/100 ml represents a critical value above which there is an increase in fetal risk (TATRA et al., 1976c)

Of the great variety of proteins produced by the human placenta, hPL has been the most widely used as an index of fetoplacental function (KUSS, 1974). Despite some conflicting reports there is general agreement on the value of hPL measurement in clinical practice, For this reason, it seemed appropriate to study the correlation between hPL and PSβG in maternal serum.

The mean curves of PSβG and hPL in normal pregnancy are similar. Consideration of paired values of PSβG and hPL from different diagnostic groups (gestosis, fetal retardation, diabetes mellitus, normals) showed a coefficient of correlation of r =0.61 (p < 0.01) (TATRA and SCHEIBER, 1978). These findings agree with those of GORDON et al. (1977), who also observed significant correlation between serum concentrations of PSβG and hPL.

CHARD (1976) has proposed four criteria which must be fulfilled by a useful indicator of placental function. PSβG seems to meet the criteria since:

1. It is not normally present in non-pregnant subjects.
2. It shows only a small day-to-day variation in individuals.
3. Its plasma half-life is relatively short.
4. The plasma concentrations are sufficiently high to allow measurement by simple and rapid techniques.

The clinical findings on maternal serum PSβG levels can be summarized as follows: Levels are elevated in twin pregnancies. In cases of diabetes and gestational diabetes there was no obvious difference from normal pregnancies. Levels were reduced in pregnancies with EPH-gestosis, the lowest values being found in the most severe cases (index 7-9).

Low levels were associated with intrauterine fetal death and intrauterine growth retardation (TATRA, 1977).

These findings suggest that maternal PSβG levels might be a valuable new index of placental function and fetal welfare, and that the test should be further explored in larger groups. Similar conclusions have been reached by TOWLER et al. (1976), and GORDON et al. (1977).

9.6 Conclusions

(1) Trophoblastic tumours. The measurement of the serum concentration of PSβG in patients with trophoblastic tumours is not as valuable as that of hCG. Only in isolated cases in which PSβG persists after hCG has become undetectable does the measurement - of PSβG give additional information.

(2) Non-Trophoblastic Tumours. The determination of PSβG serum concentration in patients with carcinoma of the breast, the intestine or the ovary does not seem to give practical information on the extension and evolution of the disease. On the other hand, detection of PSβG in carcinoma tissue itself may have prognostic significance.

(3) Early Pregnancy. In early pregnancy, measurement of PSβG could be an additional parameter for identifying cases of missed abortion or anembryonic pregnancy.

(4) Amniotic Fluid. Measurement of PSβG concentration in amniotic fluid may be a valuable clinical index of a high-risk fetus.

(5) Second Half of Pregnancy. Maternal serum concentration of PSβG seems to be a useful indicator in EPH-gestosis and fetal growth retardation. More extensive studies are needed, but should the earlier studies be confirmed there is no doubt that estimation of PSβG has much to offer in terms of the simplicity, rapidity and precision of the technique of measurement.

References

Bagshawe, K.D.: Choriocarcinoma - The clinical biology of the trophoblast and its tumours. London: Arnold 1969

Bohn, H.: Detection and characterization of pregnancy proteins in the human placenta and their quantitative immunochemical determination in sera from pregnant women. Arch. Gynäkol. 210, 440-457 (1971)

Bohn, H.: Studies on the pregnancy-specific-$\beta_1$glycoprotein (SP-1). Arch. Gynäkol. 216, 347-358 (1974)

Bohn, W., Weinmann,: Immunological disruption of implantation in monkeys with antibodies to human pregnancy-specific-$\beta_1$glycoprotein. Arch. Gynäkol. 217, 209-218 (1974)

Bohn, H., Weinmann, E.: Antifertility effect of an active immunization of monkeys with human pregnancy-specific-$\beta_1$glycoprotein. Arch. Gynäkol. 221, 305-312 (1976)

Braunstein, G.D., Vaitukaitis, J.L., Carbone, P.P., Ross, G.T.: Ectopic production of human chorionic gonadotrophin by neoplasms. Ann. Intern. Med. 78, 39-45 (1973)

Cerni, C., Tatra, G., Bohn, H.: Immunosuppression by human placental lactogen (HPL) and the pregnancy-specific-$\beta_1$glycoprotein (SP-1). Arch. Gynäkol. 223, 1-7 (1977)

Chard, T.: The normal population. In: Plasma hormone assays in the assessment of fetoplacental function. KLOPPER, A. (ed.), pp. 5-13, Edinburgh: Churchill-Livingstone 1976

Gordon, Y.B., Grudzinskas, J.G., Lewis, J.D., Jeffrey, D., Letchworth, A.T.: Circulating levels of pregnancy-specific-$\beta_1$glycoprotein and human placental lactogen in the third trimester of pregnancy: their relationship to parity, birth-weight and placental-weight. Br. J. Obstet. Gynaecol. 84, 642-647 (1977)

Grudzinskas, J.G., Gordon, Y.B., Jeffrey, D., Chard, T.: Specific and sensitive determination of pregnancy-specific-$\beta_1$glycoprotein by radioimmunoassay. A new pregnancy test. Lancet I, 333-340 (1977a)

Grudzinskas, J.G., Lenton, E.A., Gordon, Y.B., Kelso, I.M., Jeffrey, D.: Circulating levels of pregnancy-specific-$\beta_1$glycoprotein in early pregnancy. Br. J. Obstet. Gynaecol. 84, 740-742 (1977b)

Horne, C.H.W., Reid, I.N., Milne, G.D.: Prognostic significance of inappropriate production of pregnancy proteins by breast cancers. Lancet VII, 279-282 (1976a)

Horne, C.H.W., Reid, I.N., Towler, C.M., Milne, G.D.: Production of pregnancy-specific-$\beta_1$glycoprotein by nontrophoblastic tumors. 24th colloquium 1976, protides of biological fluids, p.567. Oxford: Pergamon Press 1976b

Horne, C.H.W., Towler, C.M., Pugh-Humphreys, R.G.P., Thomson, A.W., Bohn, H.: Pregnancy-specific-$\beta_1$glycoprotein - a product of the syncytiotrophoblast. Experientia 32, 1197-1199 (1976c)

Johannsen, R., Haupt, H., Bohn, H., Heide, K., Seiler, F.R., Schwick, H.G.: Inhibition of the mixed leucocyte culture (MLC) by proteins: mechanism and specifity of the reaction. Z. Immunitaetsforsch. 152, 280-286 (1976)

Kuss, E.: Klinisch-chemische Untersuchungen zur Überwachung der gefährdeten Schwangerschaft. Gynäkologie 7, 124-150 (1974)

Lin, T.M., Halbert, S.P.: Placental localization of human pregnancy-associated plasma proteins. Science (N.Y.) 193, 1249-1252 (1976)

Searle, F., Leake, B.A., Bagshawe, K.D., Dent, J.: Serum-SP-1-pregnancy-specific-$\beta_1$glycoprotein in choriocarcinoma and other neoplastic diseases. Lancet 8064 I, 579-580 (1978)

Sedlacek, H.H., Rehkopf, R., Bohn, H.: Immunofluorescence histological localization of human pregnancy and placental proteins in the placenta of man and monkey (Cynomolgus). Behring Inst. Mitt. 57, 42-49 (1976)

Sørensen, S., Borggaard, B., Rolff, L.: A radioimmunoassay of the pregnancy-specific-$\beta_1$glycoprotein (SP-1). Scand. J. Clin. Lab. Invest. 37, 537-543 (1977)

Tatarinov, Y.S., Sokolov, A.V.: Development of a radioimmunoassay for pregnancy-specific-$\beta_1$globulin and its measurement in serum of patients with trophoblastic and nontrophoblastic tumors. Int. J. Cancer 19, 161-165 (1977)

Tatarinov, Y.S., Falaleeva, D.M., Kalashnikov, V.U., Toloknov, B.O.: Immunofluorescent localization of human pregnancy-specific-βglobulin in placenta and chorioepithelioma. Nature 260, 263-267 (1976)

Tatra, G., Breitenecker, G., Gruber, W.: Serum concentration of pregnancy-specific-$\beta_1$glycoprotein (SP-1) in normal and pathologic pregnancies. Arch. Gynäkol. 217, 383-390 (1974)

Tatra, G., Placheta, P., Breitenecker, G.: Pregnancy-specific-β_1-glycoprotein: clinical aspects. Wien. Klin. Wochenschr. 87, 279-281 (1975)

Tatra, G., Tempfer, H., Placheta, P.: Influence of blood glucose levels on serum concentrations of pregnancy specific protein SP-1 and HPL during the last trimester of pregnancy, Eur. J. Obstet. Gynecol. Reprod. Biol. 6, 53-58 (1976a)

Tatra, G., Tempfer, H., Gruber, W.: Influence of 16-β-methyl-prednisone on serum concentrations of pregnancy specific protein SP-1 and HPL during the last trimester of pregnancy. Eur. J. Obstet. Gynecol. Reprod. Biol. 6, 59-62 (1976b)

Tatra, G., Polak, S., Placheta, P.: Concentration of pregnancy-specific-β_1 protein SP-1 in amniotic fluid in normal and pathologic pregnancies. Arch. Gynäkol. 221, 161-166 (1976c)

Tatra, G., Stopfer, H., Scheiber, V.: Serum concentration of SP-1, HPL and Estriol in last trimester of normal pregnancy: Circadian variations and Influence of labor. Z. Geburtshilfe Perinatol. 180, 215-219 (1976d)

Tatra, G.: Pregnancy-specific-$\beta_1$glycoprotein (SP-1) in the second half of pregnancy. Problems in perinatal medicine No. 4. Wien, München, Bern: Maudrich 1977

Tatra, G., Scheiber, V.: Correlation of maternal serum concentration of hPL and SP-1 in the second half of pregnancy. Z. Geburtshilfe Perinatol. 182 (1978) (in press)

Tatra, G., Tempfer, H., Placheta, P.: Influence of amino acid administration on serum concentrations of pregnancy-specific proteins SP-1 and hPL during last trimester of pregnancy. Eur. J. Obstet. Gynecol. Reprod. Biol. 7, 29-32 (1977)

Towler, C.M., Horne, C.H.W., Jandial, V., Campbell, D.M., MacGillivray, I.: Plasma levels of pregnancy-specific-$\beta_1$glycoprotein in normal pregnancy. Br. J. Obstet. Gynaecol. $\underline{83}$, 775-779 (1976)

Weintraub, B.D., Rosen, S.W.: Ectopic production of human chorionic somato-mammotrophin by nontrophoblastic cancers. J. Clin. Endocrinol. $\underline{32}$, 94-101 (1971)

10 Practical and Theoretical Considerations in the Measurement of Pregnancy-Specific β_1 Glycoprotein

C.H.W. Horne, R.D. Bremner, V. Jandial, R.G. Glover, and C.M. Towler

Pregnancy-specific β-globulins have been independently identified by TATARINOV and MASYUKEVICH (1970), BOHN (1971) and LIN et al. (1974). Although they have been given a variety of names, e.g. trophoblast-specific beta-globulin (TBG), schwangerschafts-spezifische β_1-glykoprotein (SP$_1$) and pregnancy-associated plasma protein (PAPP-C), it is now clear that each group had described the same protein. In this paper we have chosen to refer to it as pregnancy-specific β_1-glycoprotein (PSβG) (TOWLER et al., 1976).

PSβG is a glycoprotein - its molecular weight has been estimated as 90,000 (BOHN, 1971) and 110,000 (LIN et al., 1974). It contains some 29% carbohydrate (BOHN, 1972) and recent work suggests

that the molecule consists of a single polypeptide chain with histidine as the N-terminal amino acid (BOHN et al., 1976; BOHN
and KRAUS, 1977). A number of workers have determined its isoelectric point, the estimates ranging from 3.9 (LIN et al.,
1974) to 4.15 (BOHN, 1976). A minor component with an isoelectric point near 6.0 has also been observed (LIN et al., 1974;
TOWLER et al., 1978). This latter feature would suggest that
there is microheterogeneity in the carbohydrate component of
the molecule. However, evidence of genetic variation in the
polypeptide chain has also been observed (TOWLER et al., 1978).

There is now a considerable body of evidence in favour of PSβG
being a product of the syncytiotrophoblast. Several groups
(BOHN and SEDLACEK, 1975; HORNE et al., 1976c; TATARINOV et al.,
1976a; LIN and HALBERT, 1976) using immunohistochemical techniques have shown that it is detectable in this cell.

PSβG is detectable in maternal plasma as early as 7 days post-
ovulation (GRUDZINSKAS et al., 1977). Its serum concentration
rises steadily until about 36 weeks of gestation and thereafter
the levels remain constant till term. Following delivery the
levels fall precipitately and by extrapolation from its rate
of disappearance, its half-life would appear to be between 30
and 40 h (BOHN, 1974; TATRA et al., 1975).

In this paper we have chosen to:
1. *Outline the various methods of assay* and to indicate the most appropriate technique for any chosen sample in relation to gestation.
2. *Delineate its possible clinical uses in:* (1) diagnosis of pregnancy;
   (ii) detection of twins; (iii) prediction of outcome of
   threatened abortion; (iv) testing placental function in pregnancy complications such as pre-eclampsia and intrauterine
   growth retardation of the fetus; (v) monitoring of patients
   with trophoblastic tumours.
3. *Draw attention to possible problems* in both the measurement of
   PSβG and in interpretation of results.

10.1 Measurement of PSβG

At present there are three methods of assay in use, namely
single radial immunodiffusion (SRID), electroimmunoassay (EIA)
and radioimmunoassay (RIA). The selection of any particular
technique largely depends on the nature of the clinical material
(Fig. 10.1).

10.1.1 Single Radial Immunodiffusion (SRID)

SRID was the first immunoassay technique to be developed (MANCINI
et al., 1965). It is simple (assay time 20-72 h) and best
suited to measuring PSβG levels in the second and third trimester of pregnancy where concentrations range from 15,000 to
150,000 μg/litre. In view of the relative insensitivity of this

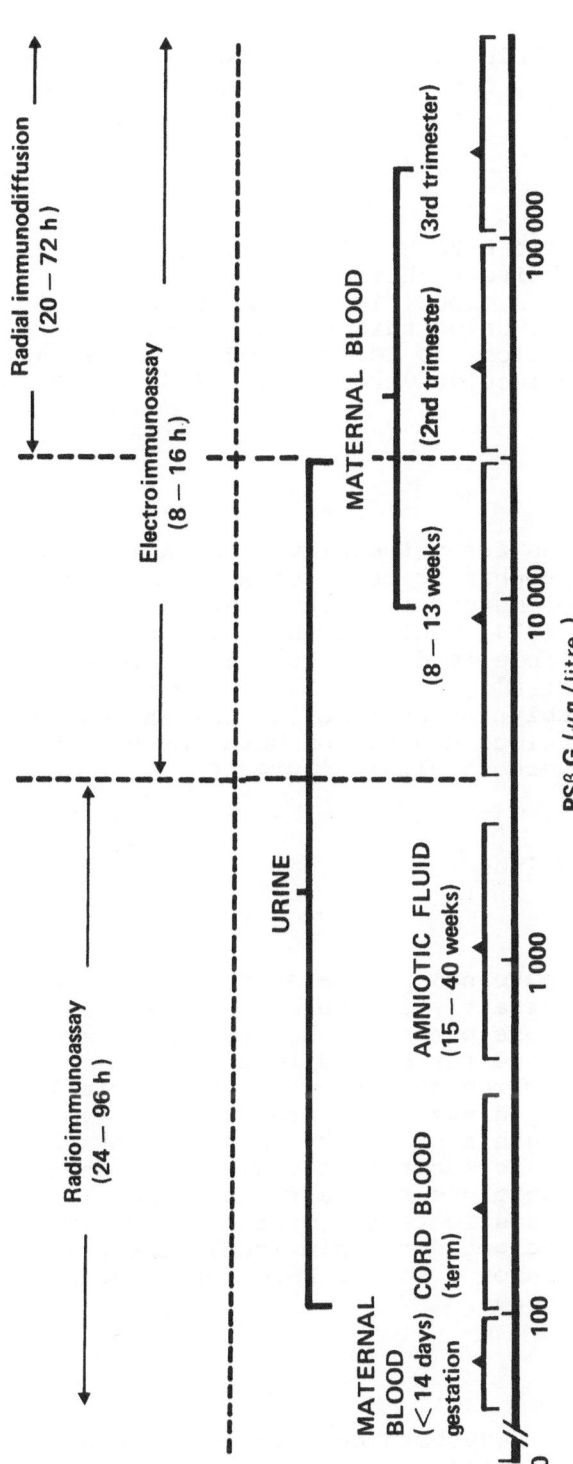

Fig. 10.1. Sensitivity range of various immunoassay techniques and PSβG values in various body fluids

145

method, however, it is not suitable for assay of PSβG in certain body fluids, e.g. amniotic fluid, cord blood. The main virtue of this technique is its simplicity.

10.1.2 Electroimmunoassay

Both simple and rapid (assay time 8-20 h) electroimmunoassay (LAURELL, 1966) is a more sensitive technique than SRID and is used by several groups of workers (BOHN, 1974; TOWLER et al., 1977a; BRUCE and KLOPPER, 1978). Using this technique it is possible to measure PSβG concentrations as low as 2000 μg/litre. An additional advantage is that molecular variants of PSβG can be detected (see Sect. 10.1.1).

10.1.3 Radioimmunoassay

By far the most sensitive of the assay techniques is radioimmunoassay though it requires more sophisticated equipment. A number of groups have described radioimmunoassay techniques for measuring PSβG (GRUDZINSKAS et al., 1977; TOWLER et al., 1977b). Using such an assay it is possible to detect PSβG at concentrations as low as 4 μg/litre (TOWLER et al., 1977b). Since the time taken for the assay varies considerably (assay time 14-96h) the decision as to whether to use this technique depends on both the nature of the specimen and whether there is clinical urgency.

10.2   Clinical Uses of PSβG Values

10.2.1 Diagnosis of Pregnancy

Since PSβG can be detected in maternal plasma shortly after implantation of the ovum it is perfectly feasible, using a sensitive radioimmunoassay, to diagnose pregnancy before the first missed period. The clinical demand for such a test would be small, for example in cases in which ovulation was artificially induced or the mother was taking drugs which are potentially teratogenic. One advantage of such a pregnancy test is that PSβG does not appear to share antigenic determinants with any other known protein or hormone and therefore the test should be absolutely specific. Current diagnostic tests, most of which involve the detection of chorionic gonadotrophin (hCG), are not specific and so the introduction of a simple PSβG latex agglutination test would be advantageous.

10.2.2 Detection of Twins

Plasma hPL and hCG measurements have both been found to be a useful adjunct to ultrasonography in the detection of twins, the levels observed being around or above the upper limit of normal (GRENNERT et al., 1976; JOVANOVIC et al., 1977). This has also been shown to be true of PSβG. In an earlier study we showed

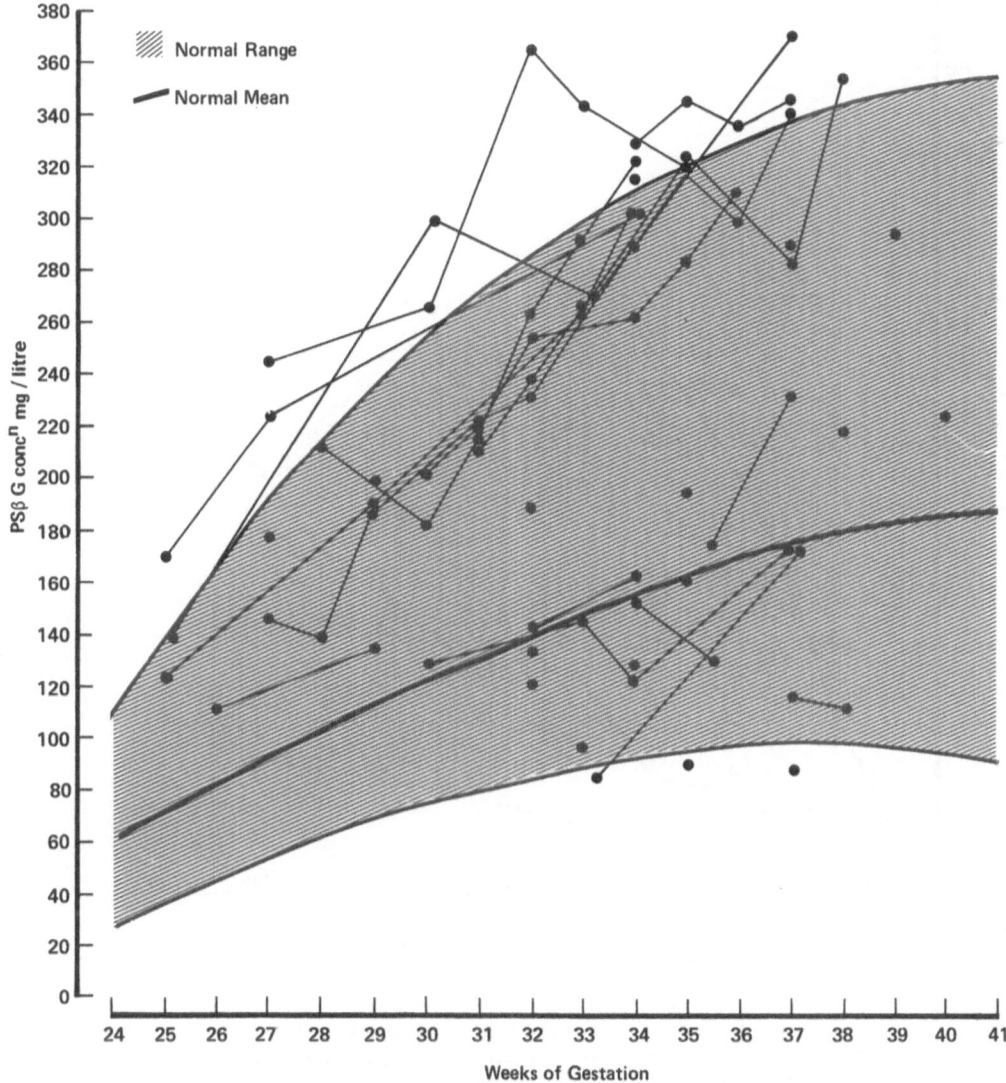

Fig. 10.2. Plasma PSβG levels in subjects with twin pregnancy (.-.) shown in relation to the normal range for singleton pregnancies (hatched area)

that in nine cases of twin pregnancy PSβG levels were around or above the upper limit (TOWLER et al., 1976). These observations have now been extended and PSβG values in a series of 42 twin pregnancies are shown in Fig. 10.2. It would appear that there are probably two distinct populations, those whose PSβG values are around or above the upper limit and a smaller group of 16 cases whose values lie close to the mean. Retrospective analysis has shown that, of this latter group, 60% were associated with growth retardation of one or both infants (the birth weights being less than the 10th centile as calculated for a singleton infant).

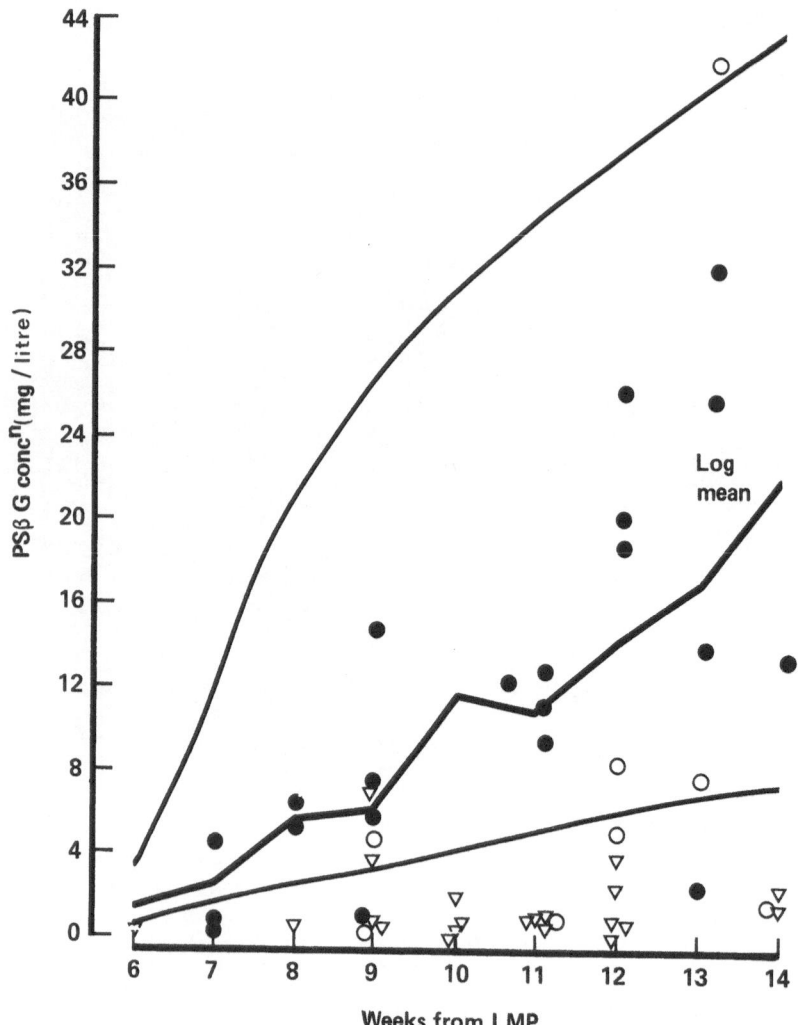

Fig. 10.3. Plasma PSβG levels (measured by radioimmunoassay) in patients presenting with threatened abortion. Mean PSβG ± 2 S.D. following logarithmic transformation of values is also shown
● Continuing pregnancies
▽ Non-viable pregnancies
○ Subsequent abortion after hospital discharge

10.2.3 Prediction of Outcome of Threatened Abortion

Measurement of the plasma levels of the trophoblast-specific proteins, placental lactogen (hPL) and hCG has been used to predict the outcome of threatened abortion (GENAZZANI et al., 1969; NIVEN et al., 1972; NYGREN et al., 1973; Gartside and TINDALL, 1975; KUNZ and KELLER, 1976) with varying degrees of success. It is not surprising therefore that the more recently described PSβG should also be used for this purpose. Work by SCHULTZ-LARSEN and HERTZ (1978) has shown that PSβG measurements are of predictive value. In a series of 59 patients these authors found that of 26 patients who aborted, 24 showed PSβG concentrations below the 95% confidence limit, a single low PSβG determination accurately predicting the inevitability of abortion in 89% of cases. With serial estimations they found the degree of accuracy increased to 100%.

Our own work supports the view that determination of PSβG concentrations are of value in predicting the outcome of threatened abortion (JANDIAL et al., 1978). Using a radioimmunoassay technique in a series of 50 cases the distribution of single PSβG estimations were as shown in Fig. 10.3. Of the 21 continuing pregnancies only 4 had PSβG levels below the 95% confidence limit whereas, of 21 cases who had non-viable pregnancies, 19 had PSβG levels below the lower limit. Of the 8 patients who aborted following discharge from hospital, however, only 4 had low PSβG measurements. While it is clear that PSβG measurements in our study are of considerable prognostic value it is likely that only serial PSβG measurements would improve the diagnostic value in those patients who aborted following hospital discharge.

10.2.4 Assessment of Fetal Growth

An important clinical use of plasma PSβG measurements is as an index of fetal growth. PSβG levels are known to correlate with fetal weight (TATRA et al., 1975; GORDON et al., 1977; TOWLER et al., 1977a) and in preliminary studies (TATRA et al., 1975; JANDIAL et al., 1977) PSβG concentrations were often low in cases in which severe preeclampsia was associated with intrauterine growth retardation (IUGR). Later work showed that low plasma PSβG levels were observed in 60%-70% of women who were delivered of infants with a birth weight below the 10th centile (GORDON et al., 1977; TOWLER et al., 1977a). In the latter reports PSβG appeared to be more useful than either hPL or oestriol in predicting IUGR.

This association has been amply confirmed in our recently completed prospective study of 100 cases in which there was clinical suspicion of a "light for dates" fetus. This opinion was based on factors such as poor maternal weight gain, antepartum haemorrhage, bad obstetric history or history of a previous delivery of a growth-retarded infant.

Figures 10.4 and 10.5 show, respectively, plasma PSβG and hPL levels in relation to birth weight. It is evident that low PSβG levels, i.e. levels more than 2 S.D. below the mean are associated with birth weights below the 10th centile (IUGR). Analysis of the data shows that low PSβG levels are present in 47% of patients who gave birth to growth-retarded infants, the sensitivity of the test (the percentage of the total patients who gave birth to infants of normal weight) being around 84%. The corresponding values for hPL are a detection rate of 32% with a sensitivity of 87%. In this study PSβG would seem to be superior to hPL in predicting IUGR.

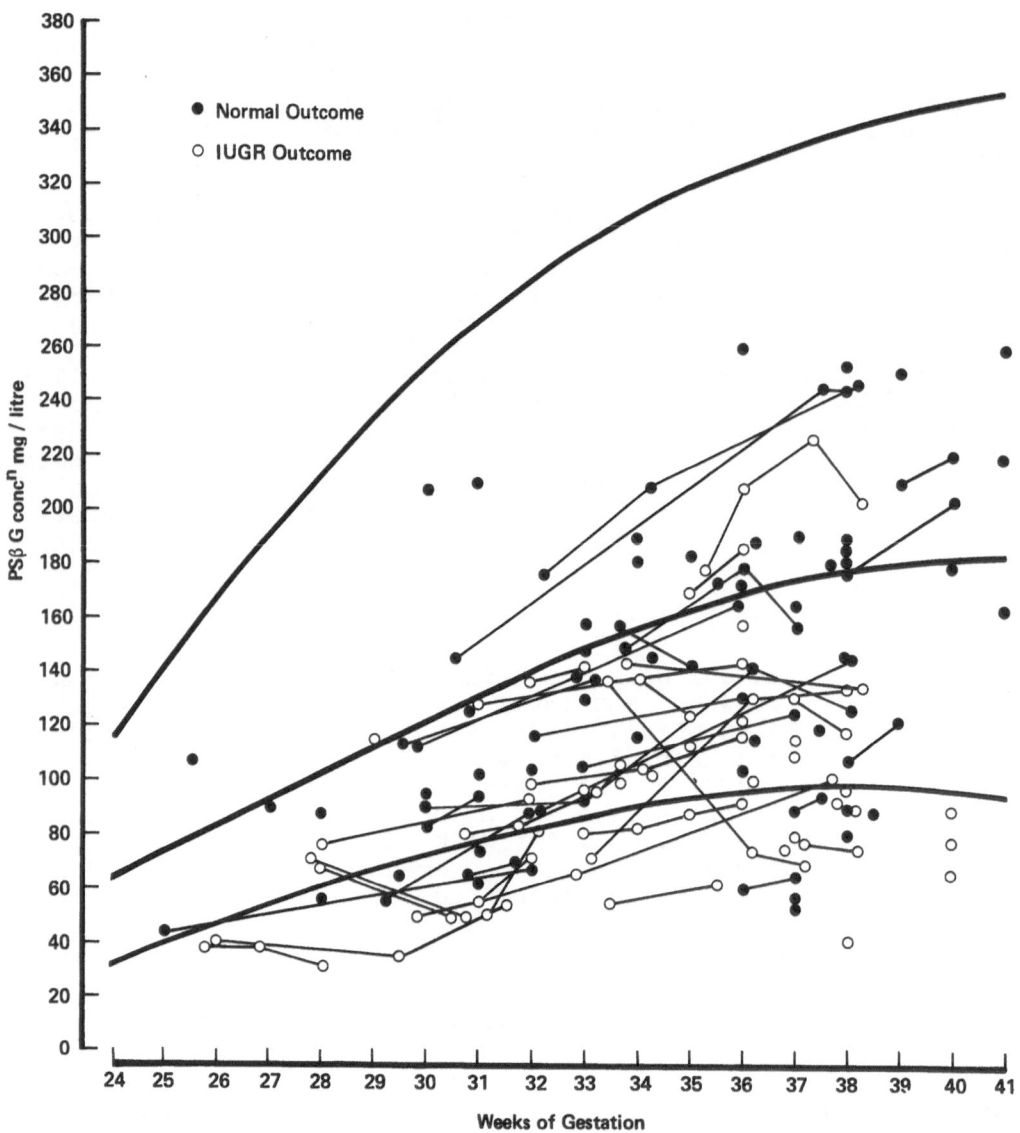

Fig. 10.4. Plasma PSβG levels in subjects who were later delivered of normal-weight infants (●) and growth-retarded infants (O). Mean and normal range also shown

Raised plasma PSβG levels have been found to indicate increased fetal size. A retrospective study (BEEBEEJAUN et al., 1978) of 20 patients who gave birth to infants over 4 kg (90% being above the 90% centile of the normal range) has shown that plasma PSβG concentrations within 30 min of delivery are higher than those of a comparable control group who gave birth to infants of normal weight (Fig. 10.6). The mean PSβG concentration of the former group (216 mg/litre) is significantly higher ($p < 0.005$) than that of the latter (143 mg/litre).

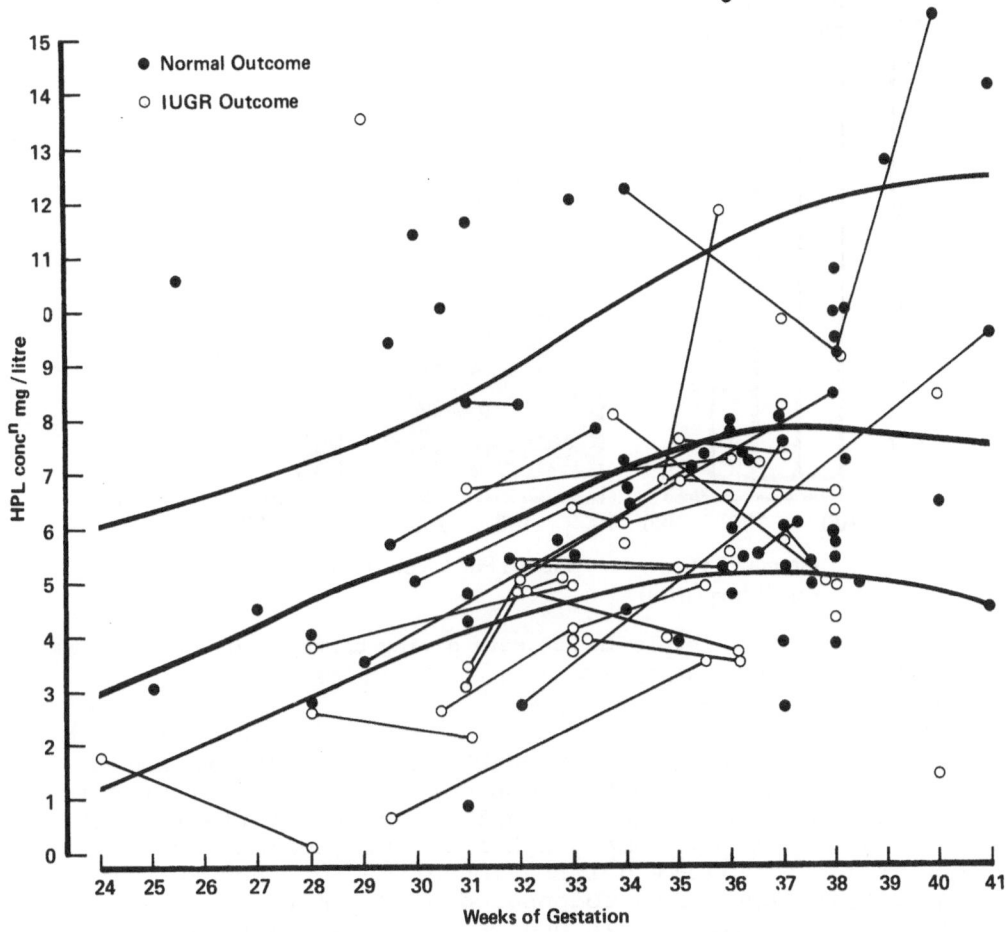

Fig. 10.5. Plasma hPL levels in same subjects as in Fig. 10.4

10.2.5 Monitoring of Trophoblastic Tumours

There are as yet few published papers on the usefulness or otherwise of PSβG in the monitoring of patients with tropho-blastic tumours. The earliest reported study appears to be that of TATARINOV et al. (1974). Using a simple double immunodif-fusion technique they were able to detect the presence of PSβG in the serum of female patients (18) with trophoblastic tumours. With the development of an immunoautoradiographic technique they were subsequently able to show that in larger series of cases a high percentage (around 80%) of female patients with various trophoblastic tumours had detectable amounts of PSβG in the peripheral blood (TATARINOV et al., 1976b;c).

With the introduction of radioimmunoassay techniques for PSβG it became possible to measure the plasma levels in female pa-tients with trophoblastic tumours even after surgery and chemo-therapy (TATARINOV and SOKOLOV, 1977). However, to date there

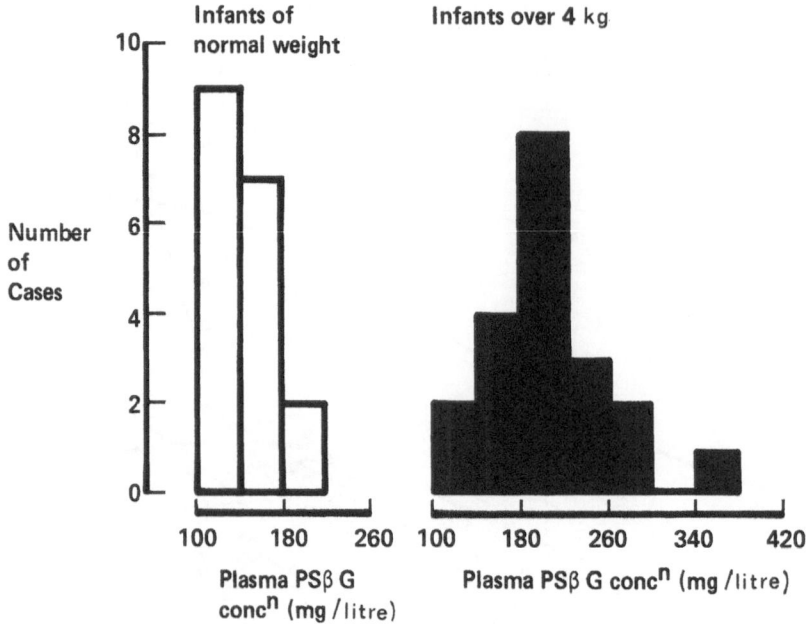

Fig. 10.6. Plasma PSβG levels in subjects giving birth to infants of normal weight and to infants weighing over 4 kg

is only one published study (SEPPÄLÄ et al., 1978) in which PSβG has been compared to the 'conventional' trophoblastic tumour marker, hCG. In a longitudinal study of 8 patients with gestational choriocarcinoma Seppälä and his colleagues were able to show that hCG persisted for a longer period than PSβG following chemotherapy. Both reappeared where there was clinical recurrence of the tumour. In one patient PSβG alone was present in the plasma. It may be, therefore, that monitoring of patients with such tumours by both markers would be worthwhile and further studies are necessary to clarify this point.

The use of PSβG radioimmunoassay for monitoring of male patients with malignant teratomata with trophoblastic elements has not as yet been fully evaluated. Its presence in the plasma of such patients has already been shown in a small number of cases (TATARINOV et al., 1976c, JOHNSON et al., 1977). In preliminary studies (LANGE et al., 1978) of patients with testicular tumours we have demonstrated that a proportion of those with non-seminomatous tumours have detectable PSβG in the plasma. HCG appears, in our study, to be more useful as a marker for such tumours, although in a few cases PSβG alone was produced.

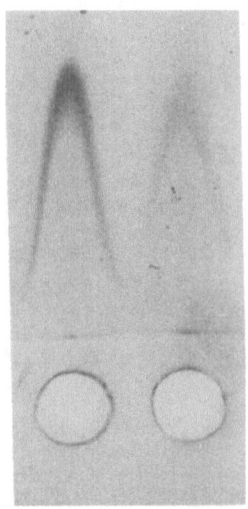

Fig. 10.7. Rocket immunoelectrophoresis of PSβG-containing sera showing normal precipitate together with indistinct variant

10.3   Problems in Measurement and Interpretation of PSβG Values

10.3.1 Variant Forms of PSβG

A study of sera from a large number of pregnant subjects has shown that some individuals produce PSβG which, while apparently in very low concentration when measured by radioimmunoassay, is normal when determined by electroimmunoassay. The "rocket" immunoprecipitates obtained with these samples in electroimmunoassay are consistently indistinct (Fig. 10.7) suggesting that they contain a form of PSβG which has an affinity for the antiserum different from that of normal PSβG. We have further demonstrated the difference between normal and abnormal PSβG using the tandem rocket technique of KROLL (1969) (Fig. 10.8). This abnormal form of PSβG is unlikely to be an artefact caused by protease digestion during sample collection, storage, preparation, or assay, since the inclusion of protease inhibitors throughout these procedures did not alter the appearance of the indistinct immunoprecipitates (TOWLER et al., 1978).

Thus it seems that a small number of patients actually produce variant forms of PSβG, although the differences between these variants and the normal protein are not yet clear; however, since neuraminidase treatment did not alter the clarity of the immunoprecipitates, a change in the carbohydrate moieties appears unlikely (TOWLER et al., 1978). It is possible, therefore, that the production of variants reflects a change in protein size or sequence.

The existence of variant forms of PSβG is an important factor in the clinical use of the immunoassay which may result in the reporting of false low levels of the glycoprotein. This is especially true in the early stages of pregnancy, when radioimmunoassay is the only technique available with sufficient sensitivity; only when PSβG levels are high enough for measurement by electroimmunoassay it is possible to distinguish between a genuinely low PSβG level and a higher level of a variant molecule.

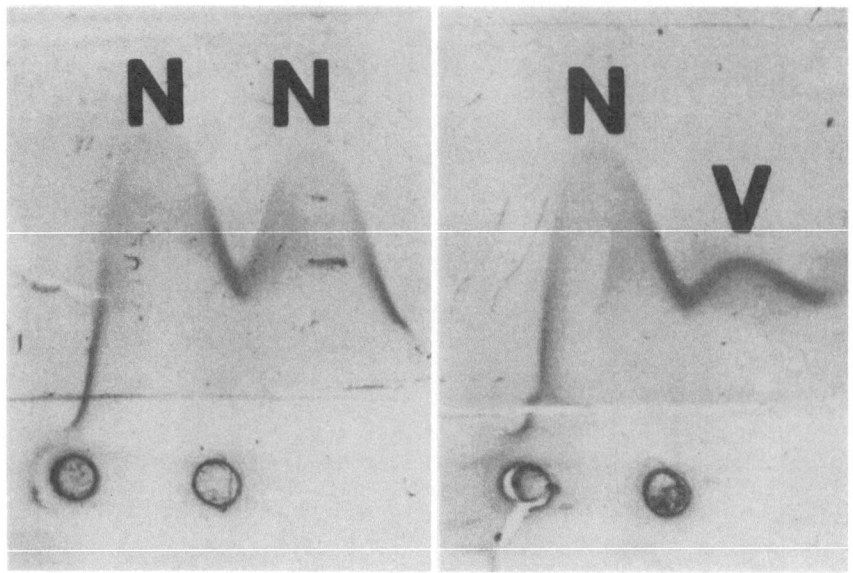

Fig. 10.8. Tandem Immunoelectrophoresis of PSβG-containing sera showing two normals (N) and normal and variant (V)

It is therefore important to know the frequency of appearance of variant forms. Of 412 subjects, variant PSβG has been detected in 18 cases (4.4%). Where serial samples have been taken from a subject producing variant PSβG, the variant form has been observed throughout the pregnancy.

Variant PSβG has been detected in both normal and abnormal pregnancies; it must be pointed out, however, that since variants cannot be identified until about the 15th week of pregnancy, it is possible that any variant form associated with early abortion would not be detected.

It is not yet clear whether all variant proteins are the same or whether a range of different variants exists. However, since the ratio of the level of PSβG as measured by radioimmunoassay to that measured by electroimmunoassay is not constant (Table 10.1), it seems possible that a number of different forms exist. The raising of antisera to variant forms should shed light on this point.

10.3.2 Production of PSβG-Like Proteins by Non-Trophoblastic Tumours

In view of the well-recognised phenomenon of ectopic production of fetal and placental proteins by tumours, attempts have been made to find evidence for PSβG production by tumour cells.

Localisation studies using indirect immunofluorescence (Fig. 10.9) and immunoperoxidase (Fig. 10.10) techniques have provided evi-

Table 10.1. Plasma PSβG concentrations as measured by radioimmunoassay (RIA) and electroimmunoassay (EIA) in pregnant subjects producing normal (n) or variant (v) PSβG

Patient	RIA (mg/litre)	EIA (mg/litre)	Immunoprecipitate (normal/variant)	Ratio RIA/EIA
L.G.	50	46	n	1.9
L.G.	50	42	n	1.19
J.H.	26	33	n	0.79
L.M.	80	118	n	0.68
J.R.	70	84	n	0.83
M.R.	18	18	n	1.00
J.S.	15	18	n	0.83
G.W.	100	99	n	1.01
G.W.	125	135	n	0.93
M.W.	102	144	n	0.71
L.B.	4.3	39	v	0.11
S.B.	1.4	22	v	0.06
A.C.	2.2	20	v	0.11
J.C.	1.4	10	v	0.14
R.C.	2.4	21	v	0.11
H.D.	16	74	v	0.21
M.F.	3.8	27	v	0.14
A.H.	3.6	32	v	0.11
C.J.	5.1	36	v	0.14
S.K.	1.8	17	v	0.11
A.L.	8.3	53	v	0.16
L.M.	3.4	36	v	0.09
V.M.	5.1	26	v	0.20
V.M.	11	48	v	0.23
P.P.	3.8	43	v	0.09
L.R.	8.9	67	v	0.13
L.S.	9.7	68	v	0.14
L.S.	7.4	52	v	0.14
S.S.	11	70	v	0.15
B.W.	6.9	51	v	0.14

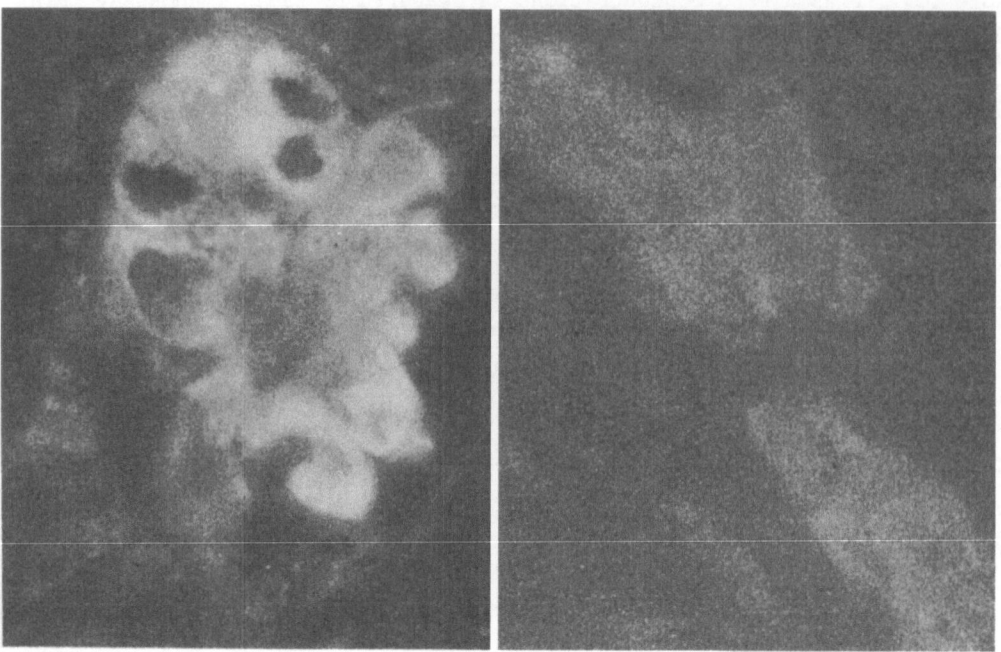

Fig. 10.9. Bronchial carcinoma stained for PSβG using an indirect immunofluorescence technique. Control (conjugate alone) shown on right (x1000)

dence for the presence of a PSβG-like protein in a number of non-trophoblastic tumours. Furthermore, a retrospective study of breast cancer patients (HORNE et al., 1976a) has revealed that there is some correlation between the presence of PSβG and the prognosis of the patient. In collaboration with Dr. W.R. Miller, University Department of Surgery, Edinburgh, we have shown that a PSβG-like protein is present in supernatants taken from a cultured mammary carcinoma line, MCF-7.

The frequency with which PSβG has been observed in tumours may well depend on the technique used to detect it. In a study of various non-trophoblastic tumours using the enzyme-bridge immunoperoxidase technique 55% of cases were found to be positive for PSβG (HORNE et al., 1976b). Lower frequencies are, however, found when sera from cancer patients are examined; in a study employing a sensitive radioimmunoassay, only 15% of sera from patients with non-trophoblastic malignancies were found to have elevated levels (3-12 µg/litre) of the glycoprotein (TATARINOV and SOKOLOV, 1977). We have recently also used radioimmunoassay to detect low levels of PSβG in sera from patients with various carcinomas; again only a small proportion (10%) of the sera contained the glycoprotein and it was also found in a smaller number of apparently normal control patients.

It is possible that the PSβG produced by tumours is an altered form analogous to the variants observed in some pregnancy sera, and that the various immunological techniques vary in their ef-

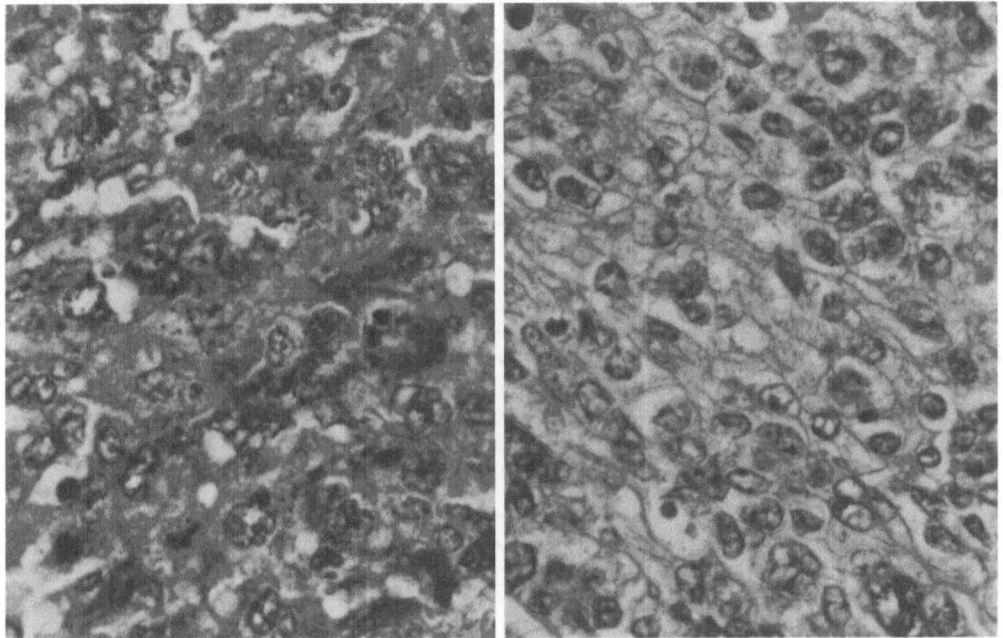

Fig. 10.10. Rectal carcinoma stained for PSβG using enzyme-bridge immunoperoxidase technique. Absorption control shown on right (x750)

ficiency of detection of such altered antigens. This would explain
the difficulty in detecting PSβG in plasma using radioimmunoassay, and would also account for the low levels observed by radioimmunoassay of homogenates of a tumour which showed strong immunofluorescent staining for PSβG.

As with the variant in pregnancy, we can only speculate on the nature of the hypothetical tumour variant. However, its behaviour
during gel filtration of crude tumour extracts suggests that it
is of a similar size to PSβG. Purification of a PSβG-like protein from a non-trophoblastic tumour is at present being attempted; it is hoped that antisera raised to this product will
facilitate further studies.

10.4 Conclusions

Assay techniques are now available for the measurement of PSβG
levels throughout pregnancy. We have attempted a critical evaluation of the value of such measurements during normal and abnormal pregnancies.

The lack of cross-reaction of PSβG with other proteins or hormones allows its use as a specific test of pregnancy; furthermore monitoring the level of PSβG during pregnancy provides an
index of the well-being of the fetus, and may be used to detect
twins, predict the outcome of threatened abortion, and detect

157

intrauterine growth retardation and increased fetal size. In
some instances (e.g. IUGR), PSβG levels have been shown to be
more useful than those of the better known pregnancy-specific
protein hPL. PSβG measurements may also prove, in conjunction
with other markers, to be of use in the monitoring of patients
with trophoblastic tumours.

We would, however, point out two problems in the interpretation
of PSβG measurements. The production of variant forms of the
glycoprotein by approximately 4% of pregnant subjects can lead
to the reporting of false low levels of PSβG, especially in the
early stages of pregnancy. The synthesis of PSβG-like proteins
by certain non-trophoblastic tumours may also interfere with
the interpretation of PSβG values; however, it is hoped that
further study of this phenomenon may yield a valuable prognostic
test for some malignant diseases.

References

Beebeejaun, M.S., Jandial, V., Towler, C.M., Sutherland, H.M., Horne, C.H.W.:
The role of pregnancy-specific β_1-glycoprotein in the management of preg-
nancies complicated by diabetes. In: Proceedings of second international
Colloquium on carbohydrate metabolism in pregnancy and the newborn. SUTHER-
LAND, H.M., STOWERS, J.M. (eds.). Berlin: Springer 1978 (in press)

Bohn, H.: Detection and characterization of pregnancy proteins in the human
placenta and their quantitative immunochemical determination in sera from
pregnant women. Arch. Gynäkol. 210, 440-457 (1971)

Bohn, H.: Isolation and characterization of pregnancy-specific β_1-glycopro-
tein. Blut 24, 292-302 (1972)

Bohn, H.: Immunochemical determination of human pregnancy proteins. Arch.
Gynäkol. 217, 219-231 (1974)

Bohn, H.: Isolation and characterization of placental specific proteins SP$_1$
and PP$_5$. Prot. Biol. Fluids 24, 117-124 (1976)

Bohn, H., Kraus, W.: Isolation and characterization of pregnancy-specific
β_1-glycoprotein from urine of pregnant women. Arch. Gynäkol. 223, 33-40
(1977)

Bohn, H., Schmidtberger, R., Zilg, H.: Isolation of pregnancy-specific β_1-
glycoprotein and antigenically related proteins by immunoadsorption.
Blut 32, 103-113 (1976)

Bohn, H., Sedlacek, H.: Comparative studies on placental specific proteins
in man and subhuman primates. Arch. Gynäkol. 220, 105-121 (1975)

Bruce, S., Klopper, A.: The measurement of pregnancy-specific β_1-glycopro-
tein by electroimmunodiffusion. Clin. Chim. Acta 84, 107-113 (1978)

Gartside, M.W., Tindall, V.R.: The prognostic value of human placental
lactogen (HPL) levels in threatened abortion. Br. J. Obstet. Gynaecol.
82, 303-309 (1975)

Genazzani, A.R., Aubert, M.L., Casoli, M., Fioretti, P., Felber, J.P.: Use
of human placental lactogen radioimmunoassay to predict outcome in cases
of threatened abortion. Lancet II, 1385-1387 (1969)

Gordon, Y.B., Jeffrey, D., Grudzinskas, J.G., Chard, T., Letchworth, A.T.:
Concentrations of pregnancy-specific β_1-glycoprotein in maternal blood
in normal pregnancy and in intrauterine growth retardation. Lancet I,
331-333 (1977)

Grennert, L., Persson, P.-H., Genser, G., Kullander, S., Thorell, J.: Ultrasound and human-placental-lactogen screening for early detection of twin
pregnancies. Lancet I, 4-6 (1967)

Grudzinskas, J.G., Gordon, Y.B., Jeffrey, D., Chard, T.: Specific and sensitive determination of pregnancy-specific β_1-glycoprotein by radioimmunoassay. Lancet I, 333-335 (1977)

Horne, C.H.W., Reid, I.N., Milne, G.D.: Prognostic significance of inappropriate production of pregnancy proteins by breast cancers. Lancet II,
279-282 (1967a)

Horne, C.H.W., Reid, I.N., Towler, C.M., Milne, G.D.: Production of pregnancy-specific β_1-glycoprotein by nontrophoblastic tumours. Prot. Biol. Fluids
24, 567-570 (1976b)

Horne, C.H.W., Towler, C.M., Pugh-Humphreys, R.G.P., Thompson, A.W., Bohn, H.:
Pregnancy specific β_1-glycoprotein - a product of the syncytiotrophoblast.
Experientia 32, 1197-1199 (1976c)

Jandial, V., Campbell, D.M., Towler, C.M., MacGillivray, I., Horne, C.H.W.:
Pregnancy specific β_1-glycoprotein - an indicator of placental function.
Br. J. Obstet. Gynaecol. 84, 76 (1977)

Jandial, V., Towler, C.M., Horne, C.H.W., Abramovich, D.R.: Plasma pregnancy-
specific β_1-glycoprotein in complications of early pregnancy. Br. J.
Obstet. Gynaecol. (1978) (in press)

Johnson, S.A.N., Grudzinskas, J.G., Gordon, Y.B., Al-Ani, A.T.M.: Pregnancy-
specific $beta_1$-glycoprotein in plasma and tissue extract in malignant teratoma of the testis. Br. Med. J. I, 951-952 (1977)

Jovanovic, L., Landesman, R., Saxena, B.B.: Screening for twin pregnancy.
Science 198, 738 (1977)

Kroll, J.: Immunochemical identification of specific precipitin lines in
quantitative immunoelectrophoresis patterns. Scand. J. Clin. Lab. Invest.
24, 55-65 (1969)

Kunz, J., Keller, P.J.: HCG, HPL, oestradiol, progesterone and AFP in serum
in patients with threatened abortion. Br. J. Obstet. Gynaecol. 83, 640-644
(1976)

Lange, P.H., Horne, C.H.W., Bremner, R.D.: Unpublished data (1978)

Laurell, C.B.: Quantitative estimation of proteins by electrophoresis in
agarose gels containing antibodies. Anal. Biochem. 15, 45-52 (1966)

Lin, T.M., Halbert, S.P.: Placental localization of human pregnancy-associated plasma proteins. Science 193, 1249-1252 (1976)

Lin, T.M., Halbert, S.P., Kiefer, D., Spellacy, W.N., Gall, S.: Characterization of four human pregnancy-associated plasma proteins. Am. J. Obstet.
Gynecol. 118, 223-235 (1974)

Mancini, G., Carbonara, A.O., Heremans, J.F.: Immunochemical quantitation
of antigens by single radial immunodiffusion. Immunochemistry 2, 235-254
(1965)

Niven, P.A.R., Landon, J., Chard, T.: Placental lactogen levels as guide to outcome of threatened abortion. Br. Med. J. III, 799-801 (1972)

Nygren, K.G., Johannson, E.D.B., Wide, L.: Evaluation of the prognosis of threatened abortion from the peripheral plasma levels of progesterone, estradiol and human chorionic gonadotrophin. Am. J. Obstet. Gynecol. 116, 916-922 (1973)

Schultz-Larsen, P., Hertz, J.: Pregnancy specific β_1-glycoprotein in threatened abortion. Scand. J. Clin. Lab. Invest. (1978) (in press)

Seppälä, M., Rutanen, E.-M., Heikinheimo, M., Jalanko, H., Engvall, E.: Detection of trophoblast tumour activity by pregnancy-specific beta-1- glycoprotein. Int. J. Cancer 21, 265-267 (1978)

Tatarinov, Yu.S., Masyukevich, V.N.: Immunochemical identification of new β_1-globulin in the blood serum of pregnant women. Byull. Eksp. Biol. Med. 69, 66-68 (1970)

Tatarinov, Yu.S., Sokolov, A.V.: Development of a radioimmunoassay for pregnancy-specific beta$_1$-globulin and its measurement in serum of patients with trophoblastic and non-trophoblastic tumours, Int. J. Cancer 19, 161-166 (1977)

Tatarinov, Yu.S., Mesnyankina, N.V., Nikoulina, D.M., Novikova, L.A., Toloknov, B.O., Falaleeva, D.M.: Immunochemical identification of beta$_1$-globulin of the "pregnancy-zone" in serum of patients with trophoblastic tumours. Int. J. Cancer 14, 548-554 (1974)

Tatarinov, Yu.S., Falaleeva, D.M., Kalashnikov, V.V., Toloknov, B.O.: Immunofluorescent localisation of human pregnancy-specific β-globulin in placenta and chorioepithelioma. Nature 260, 263 (1976a)

Tatarinov, Yu.S., Falaleeva, D.M., Kalashnikov, V.V.: Human pregnancy-specific beta$_1$-globulin and its relation to chorionepithelioma. Int. J. Cancer 17, 626-632 (1976b)

Tatarinov, Yu.S., Falaleeva, D.M., Kozljaeva, G.A., Nikoulina, D.M.: Human trophoblastic beta$_1$-globulin and chorionepithelioma. In: Oncodevelopmental gene expression. FISHMAN, W.H., SELL, S. (eds.). New York: Academic Press 1976

Tatra, G., Breitenecker, G., Gruber, W.: Serum concentration of pregnancy specific β_1-glycoprotein (SP-1) in normal and pathological pregnancies. Arch. Gynäkol. 217, 383-390 (1974)

Tatra, G., Placheta, P., Breitenecker, G.: Pregnancy-specific β_1-glycoprotein (SP$_1$): Clinical aspects. Wien. Klin. Wochenschr. 87, 279-281 (1975)

Towler, C.M., Horne, C.H.W., Jandial, V., Campbell, D.M., MacGillivray, I.: Plasma levels of pregnancy-specific β_1-glycoprotein in normal pregnancy. Br. J. Obstet. Gynaecol. 83, 775-779 (1976)

Towler, C.M., Horne, C.H.W., Jandial, V., Campbell, D.M., MacGillivray, I.: Plasma levels of pregnancy-specific β_1-glycoprotein in complicated pregnancies. Br. J. Obstet. Gynaecol. 84, 258-263 (1977a)

Towler, C.M., Horne, C.H.W., Jandial, V., Chesworth, J.M.: A simple sensitive radioimmunoassay for pregnancy-specific β_1-glycoprotein. Br. J. Obstet. Gynaecol. 84, 580-584 (1977b)

Towler, C.M., Glover, R.G., Horne, C.H.W.: Problems encountered in the measurement of pregnancy-specific β_1-glycoprotein. Clin. Chim. Acta (1978) (in press)

11 Trophoblast-Specific β_1 Glycoprotein as an Indicator of Pregnancy and Neoplasia – A Review of Recent Clinical Studies

Y. S. Tatarinov

The measurements of placental proteins and embryo-specific antigens as tests of placental function, as means of detecting pregnancy or as tumour markers are presently under intensive investigation. Thus α-fetoprotein serves as a specific marker for chemically induced mouse hepatomas (ABELEV et al., 1963) and for human hepatocellular carcinoma (TATARINOV, 1963). Carcinoembryonic antigen is used as an indicator of colonic adenocarcinoma (GOLD and FREEDMANN, 1965). These proteins are present in normal adult serum at levels from 2-40 ng/ml (EGAN et al., 1972; MCINTIRE et al., 1975; RUOSLAHTI and SEPPÄLÄ, 1971). On the other hand placental proteins cannot usually be detected in the sera of normal non-pregnant adults even by sensitive radioimmunoassay techniques. Thus placental proteins are specific indicators of the presence of either pregnancy or cancer.

In the present paper, recent clinical and experimental data concerning pregnancy-specific β glycoprotein or trophoblast-specific β_1-glycoprotein (TSG) will be reviewed.

11.1 TSG and Its Identification

TSG was first discovered in sera from pregnant women (TATARINOV and MASYUKEVICH, 1970) and in placental extracts (BOHN, 1971; TATARINOV et al., 1971). A similar protein was found in pregnancy plasma (LIN et al., 1974a). At the present time there are many different names for this protein: pregnancy-beta$_1$-globulin or PBG (TATARINOV and MASYUKEVICH, 1970); pregnancy-specific beta$_1$-glycoprotein or SP-1 (BOHN, 1971), pregnancy-associated plasma protein-C or PAPP-C (LIN et al., 1974 a,b,c), pregnancy-specific β_1 glycoprotein or PSβ_1G (TOWLER et al., 1976), trophoblast-specific beta$_1$-glycoprotein or TSG (TATARINOV and SOKOLOV, 1977) and others.

Detailed immunochemical identifications of four human pregnancy-associated proteins were carried out by LIN and HALBERT in 1975. It was shown that the so-called PAPP-C and SP-1 are identical proteins. A TSG preparation from pooled pregnancy sera, when tested against four different monospecific antisera in our laboratory, gave similar results to the findings in Bohn's and in Halbert's laboratories (TATARINOV et al., 1976e). Analogous results were demonstrated with our antisera to TSG by Halbert in 1976. These collaborative studies have confirmed that the TSG of TATARINOV and MASYUKEVICH (1970), the SP-1 of BOHN (1971) and the PAPP-C of LIN et al. (1974b) are all the same protein.

Sensitive and specific immunochemical methods are now in common use which serve for routine measurements of TSG and its experimental identification. TSG has a characteristic antigenic structure and shows no immunochemical cross-reaction with the known normal plasma proteins or with the known placental enzymes (GALL and HALBERT, 1972; TATARINOV et al., 1974c).

11.2 Characteristics of TSG

TSG was isolated and purified from placental extracts (BOHN, 1972), pooled pregnancy plasma (LIN et al., 1974b) and pooled pregnancy serum (TATARINOV and SOKOLOV, 1977) by a combination of methods including ion-exchange chromatography, gel filtration, preparative electrophoresis and isoelectric focusing. TSG was also prepared by immunoabsorbent chromatography (BOHN et al., 1976), and it was obtained in a highly purified form using this technique.

TSG is a typical glycoprotein, containing about 28% carbohydrate, which includes hexose, N-acetylhexosamine, N-acetylneuraminic acid and fucose (BOHN, 1974). The molecular weight of TSG estimated by gel filtration and gel chromatography was found to be between 110,000 (LIN et al., 1974b) and 113,000 (TATARINOV et al., 1974c). Ultracentrifugation showed a sedimentation constant of 4.6 S and a molecular weight of 120,000 (BOHN, 1972). This contrasts with the results obtained by gel electrophoresis in the presence of sodium dodecyl sulphate which gave a molecular weight of 90,000 (BOHN, 1974). TSG was also demonstrated in the urine of pregnant women (BOHN and KRAUS, 1977). This component has a molecular weight of 65,000 with a sedimentation coefficient 2.9 S.

Two isoelectric points have been recorded, one at a pH of 3.6 and the other at 6.5 (LIN et al., 1974a; NIKULINA et al., 1977). This heterogeneity of TSG is also evident on disc electrophoresis combined with double immunodiffusion (TATARINOV et al., 1976e). It is probable that this heterogeneity is associated with differences in the carbohydrate content of TSG molecules.

11.3 Function of TSG

The biological function of TSG is not yet established. TSG has an affinity for some steroid hormones (BOHN, 1974; BOHN and KRANZ, 1973) and is said to have iron-binding capacity (LIN et al., 1974a). It is important to distinguish between TSG and sex-hormone-binding globulin, which binds steroid hormones much more strongly than TSG. There are in vitro studies on the immunosuppressive activity of TSG. In mixed human lymphocyte culture it has been shown to have certain inhibitory effects (BOHN et al., 1976; CERNI et al., 1977; HORNE et al., 1976b). These findings may have an important bearing on the suppression of immunological rejection of the fetus by the mother.

11.4 Site of Synthesis

The site of TSG synthesis was studied by two experimental techniques. Firstly, the synthesis of TSG by the immature human placenta has been demonstrated, using tissue culture (HORNE et al., 1976b; TATARINOV et al., 1976b, 1977). The incorporation of ^{14}C-amino acids into TSG was demonstrated by immunoradioautography. On the other hand organs such as the liver, lung, gastrointestinal tract or muscle and endocrine glands obtained from pregnant women did not incorporate labelled amino acids under identical conditions. Secondly, by indirect immunofluorescence techniques, it has been shown that TSG was present in the trophoblastic cells of the chorion (AFANASJEVA et al., 1976, BOHN, 1975; LIN and HALBERT, 1976; TATARINOV et al., 1975a). Specific fluorescence was found in the cytotrophoblast and syncytiotrophoblast of the chorion in immature placentae and also in residual trophoblast and in fibrin deposits in mature placentae. The latter may, of course, have contained necrotic trophoblast cells. Using an enzyme-bridge immunoperoxidase method, HORNE et al. (1976b) demonstrated TSG in the endoplasmic reticulum and in the membranes of the syncytiotrophoblast.

It is uncertain whether TSG is produced by the cytotrophoblastic cells only, or by the syncytiotrophoblast as well. Electron microscopic investigation (WISLOCKI and DEMPSEY, 1955) of the structure of the human placenta has revealed that syncytiotrophoblastic cells are derived by differentiation from the cytotrophoblastic cells. In addition, the immunochemical localization of human chorionic gonadotrophin suggests that it is probably in both cell types (MIDGLEY and PIERCE, 1962). In this connection it can be suggested that TSG production begins in the cytotrophoblastic cells, albeit at a lesser rate.

11.5 TSG Assay as a Test for Pregnancy

Using an immunodiffusion technique, TATARINOV and MASYUKEVICH (1970) were the first to demonstrate TSG in 97.4% of pregnant women at 6-8 weeks and in 100% at later stages of gestation. These authors suggested that TSG determination in the serum of pregnant women could be used for the diagnosis of pregnancy. Subsequently, the immunodiffusion test was employed for the early diagnosis of gestation (TATARINOV et al., 1974c). By radio-immunoassay, TSG was detected as early as 18-23 days after ovulation (GRUDZINSKAS et al., 1977). These results suggest that radioimmunoassay may detect TSG very early in pregnancy, and have been confirmed by subsequent studies (LIN et al., 1976; TOWLER et al., 1976).

In routine applications electroimmunodiffusion or latex agglutination methods for TSG assay have a sensitivity of 100 ng/ml, which is sufficient to make them a valid alternative to radioimmunoassay.

11.6 TSG Assay as a Fetoplacental Function Test

After implantation of the ovum, maternal TSG levels correlate with the stage of pregnancy and the measurement of TSG in maternal serum may be employed for the detection of fetoplacental well-being (TOWLER et al., 1976). TATRA et al. (1976) have reported that TSG is correlated with newborn weight and with placental weight but LIN et al., (1976) could not confirm this correlation.

Serial determinations of serum TSG showed a fall in TSG levels in some diseases of pregnancy (ABRAMOVA and TATARINOV, 1978; AFANASJEVA et al., 1976; GORDON et al., 1977; TATRA et al., 1974). The TSG concentration in pre-eclamptic toxaemia is approximately half that of normal pregnancy (AFANASJEVA et al., 1976); In eclampsia, TSG levels were often below normal (TATRA et al., 1974). In cases of intrauterine fetal death, the TSG levels may remain unchanged for several days or fall as rapidly as after normal delivery (ABRAMOVA and TATARINOV, 1978; TATRA et al., 1974; TATRA et al., 1975). Several preliminary observations suggest that measurement of TSG in maternal serum and amniotic fluid could be useful for monitoring normal and pathological pregnancy and could provide some information concerning intrauterine growth retardation (JANDIAL et al., 1977; TOWLER et al., 1977; TATRA et al., 1976; GORDON et al., 1977).

11.7 TSG Test in Malignancy

11.7.1 Trophoblastic Tumours

Using an immunodiffusion technique, TATARINOV et al. (1974) were the first to show TSG in the serum of 30% of patients with chorionepithelioma and other trophoblastic tumours. Subsequently, TSG was detected in some patients with testicular teratoblastomas (JOHNSON et al., 1977; TATARINOV et al., 1975b) by immunoradioautography in agar or by double-antibody radioimmunoassay. These findings may open new vistas in the study of neoplasia.

During the last 6 years, we have used (TATARINOV et al., 1974a, b, 1975b, 1976a,c,d; KALASHNIKOV et al., 1978) TSG assay in routine testing of a variety of pathological sera including both trophoblastic and non-trophoblastic tumours. Four immunochemical methods were used: (a) immunodiffusion in agar (sensitivity 2000-3000 ng/ml), (b) immunoradioautography in agar (sensitivity 50-100 ng/ml), (c) double antibody radioimmunoassay (sensitivity 1-3 ng/ml), and (d) immunoenzymeassay (sensitivity 3-6 ng/ml). It was found that significantly more positive tests for TSG occurred with immunoradioautography (76.6%) than with immunodiffusion (33.3%) when sera from patients with post-delivery tumours and chorioepitheliomas were examined (TATARINOV et al., 1976a). It is interesting that the immunoradioautographic test for TSG was positive both before treatment (82.4%) and after surgery and/or chemotherapy (37.2%) in patients with chorioepithelioma of the uterus. It must be assumed that TSG-producing cells continued the synthesis and the secretion of this protein into the circulation. Elevated TSG levels were also found in treated patients (76.7%) by means of radioimmunoassay (TATARINOV and SOKOLOV, 1977). Specific studies will be required to ascertain the prognostic significance of this observation.

By radioimmunoassay and immunoenzymeassay, elevated TSG levels can be demonstrated in the majority of patients with untreated chorioepithelioma as well as with post-delivery or post-molar trophoblastic tumours (SEPPÄLÄ et al., 1978; KALASHNIKOV et al., 1978).

The cellular localization of TSG was studied by indirect immunofluorescence (TATARINOV et al., 1976a,b). TSG was localized in certain cells in bulky masses or in nodules of the chorioepithelioma. Specific fluorescence was also associated with the membrane of giant trophoblastic cells of chorioepithelioma (TATARINOV et al., 1976a).

The relationship between TSG and human chorionic gonadotrophin (hCG) in trophoblastic tumours was first studied by immunoradiography for serum TSG and by a haemagglutination test for urinary hCG (TATARINOV et al., 1976d). Subsequently, radioimmunoassays for TSG and for hCG β subunit were done in serum (JOHNSON et al., 1977; BAGSHAWE et al., 1978; SEPPÄLÄ et al., 1978). It was shown that there is a correlation between serum TSG and urinary and/or serum hCG levels. However, some dissociation of serum TSG and serum hCG levels was found in gestational chorioepitheliomas and especially in patients with malignant teratomas. Apparently, this dissociation is due to differences in the synthesis and metabolism of TSG and hCG, although these proteins are produced by the same cell in both normal and malignant trophoblast. Another explanation may be that both markers are synthesized within the same cell but that the biosynthetic process is differentially switched on or off during the growth of the trophoblast cell

(SEPPÄLÄ et al., 1978). Further studies will be required to determine whether TSG measurements might detect minimal residual
trophoblastic disease, when hCG measurements are negative.

Thus the high sensitivity and specificity of TSG assays can be
used not only for the diagnosis and evaluation of the effectiveness of surgical or chemotherapeutic treatment, but these assays
might also have an important potential in epidemiological investigations, particularly in population groups with a high risk
of trophoblastic disease, as, for instance, in women who have
had a molar pregnancy.

11.7.2 Non-Trophoblastic Tumours

Tissue TSG was detected in 60% of breast cancers and in 50% of
gastrointestinal tumours by means of an enzyme-bridge immunoperoxidase technique (HORNE et al., 1976a,c). By radioimmunoassay in plasma, however, elevated TSG levels in non-trophoblastic tumours could be demonstrated in very few cases (11 out
of 114 patients). In general, the TSG in such cases is low,
3-12ng/ml. Only 2 of 114 patients with lung carcinoma and tera-
toblastoma of the mediastinum had TSG levels of around 25 ng/ ml
(TATARINOV and SOKOLOV, 1977).

The presence of TSG in some non-trophoblastic malignancies may
be due to the activation of a gene which is suppressed in differentiated adult tissues. However, this hypothesis remains to
be substantiated, because the ectopic production of TSG is not
necessarily similar to the ectopic production of hCG. Indeed the
ectopic production and secretion of hCG is relatively common.
As was reported by ROSEN et al. (1975), a high percentage of
some non-trophoblastic tumours such as testicular carcinomas
(51%), ovarian carcinomas (42%) gastrointestinal tumours (17%)
and lung carcinomas (9%) secrete hCG into the circulation.

The origin of TSG synthesis by non-trophoblastic tumours requires further investigation.

11.8 Conclusions

The discovery of TSG in the sera of pregnant women and cancer
patients, more particularly in patients with post-delivery and
postmolar trophoblastic tumours, opens new prospects in the
study of pregnancy-associated and tumour-associated protein
markers. At present, maternal serum TSG deserves consideration
as a parameter for the diagnosis of early gestation as well as
a valuable indicator for the monitoring of normal and pathological pregnancy. The breakdown of tolerance to homologous TSG
should be very promising for the development of a specific contraceptive vaccine. However, all problems associated with homologous immunization against chemically modified TSG have not
yet been solved. An experimental model (LIN et al., 1975; TATA-
RINOV et al., 1976e; SHEVCHENKO et al., 1977) would be helpful
for the investigation of the role of TSG in pregnancy and in
trophoblastic and non-trophoblastic tumours.

The development of a reliable immunochemical test for TSG in the diagnosis and the monitoring of chorioepithelioma and other trophoblastic tumours seems to be most important. TSG has a characteristic antigenic structure, since this protein shows no immunochemical cross-reaction with the known placental hormones and placental enzymes. Using the TSG test, there will be no false positives in patients treated with pituitary or chorionic gonadotrophin. In this case, the measurement of TSG may be the only sure evidence of trophoblastic activity both in pregnancy and malignant disease.

Finally, it is necessary to have international standard preparations of human and animal TSG for the comparison of the results obtained by different workers on the physicochemical, biochemical, biological and physiological properties of the protein as well as for comparison in clinical studies.

References

Abelev, G.I.: Alpha-fetoprotein in ontogenesis and its association with malignant tumours. Adv. Cancer Res. $\underline{14}$, 295-358 (1971)

Abelev, G.I.: Alpha-fetoprotein as a marker of embryo-specific differentiations in normal and tumour tissue. Transplant. Rev. $\underline{20}$, 3-37 (1974)

Abelev, G.I., Perova, S.D., Khramkova, N.I., Postnikova, Z.A., Irlin, I.S.: Embryonal serum alpha-globulin and its synthesis by the transplantable mouse hepatomas. Biokhimia, USSR, 625-634 (1963); Transplant. Bull. 1, 174-180 (1963)

Abramova, L.A., Tatarinov, Y.S.: Immunodiffusion test for trophoblastic $beta_1$-glycoprotein in the risk of abortus. Akush. Ginekol. (Mosk.) (1978) (in press)

Afanasjeva, A.V., Volodin, M.A., Nikulina, D.M., Mesnjankina, N.V., Parfenova, L.A.: A study of $beta_1$-globulin production under conditions of normally developing pregnancy and in late toxemia. Akush. Ginekol. (Mosk.) $\underline{6}$, 22-24 (1976)

Bagshawe, K.D., Lequin, R.M., Sizaret, P., Tatarinov, Y.S.: Pregnancy $beta_1$-glycoprotein and chorionic gonadotrophin in the serum of patients with trophoblastic and non-trophoblastic tumours. Eur. J. Cancer (1978) (in press)

Bohn, H.: Nachweis und Charakterisierung von Schwangerschaftsprotein in der menschlichen Plazenta, sowie ihre quantitative immunologische Bestimmung im Serum schwangerer Frauen. Arch. Gynaekol. $\underline{210}$, 440-457 (1971)

Bohn, H.: Isolierung und Charakterisierung des schwangerschafts-spezifischen $beta_1$-glycoproteins. Blut $\underline{24}$, 292-302 (1972)

Bohn, H.: Untersuchungen über das schwangerschaftsspezifische $beta_1$-Glycoprotein (SP-1). Arch. Gynaekol. $\underline{216}$, 347-358 (1974)

Bohn, H.: The protein antigens of human placenta as a basis for the development of contraceptive vaccine. 3rd int. symp. immunol. of reprod. WHO session, pp. 111-125. Varna 1975

Bohn, H., Kranz, T.: Untersuchungen über die Bindung von Steroidhormonen an menschliche Schwangerschaftsproteine. I. Identifizierung des schwangerschaftsassoziierten $beta_1$-glycoprotein mit dem Steroid-bindenden $beta_1$-globulin. Arch. Gynaekol. 215, 63-71 (1973)

Bohn, H., Kraus, W.: Isolation and characterization of pregnancy-specific $beta_1$-glycoprotein from urine of pregnant women. Arch. Gynäkol. 223, 33-39 (1977)

Bohn, H., Schmidtberger, R., Zilg, H.: Isolierung des schwangerschaftsspezifischen $beta_1$-Glykoproteins (SP-1) und antigenverwandter Proteine durch Immunoadsorption. Blut 32, 103-113 (1976)

Cerni, C., Tatra, G., Bohn, H.: Immunosuppression by human placental lactogen (HPL) and the pregnancy-specific $beta_1$-glycoprotein (SP-1). Arch. Gynäkol. 223, 1-7 (1977)

Egan, M.D., Lautenschleger, J.T., Coligan, J.E., Todd, C.W.: Radioimmunoassay of carcinoembryonic antigen. Immunochemistry 9, 289-299 (1972)

Gall, S.A., Halbert, S.P.: Antigenic constituents in pregnancy plasma which are undetectable in normal nonpregnant female or male plasma. Int. Arch. Allergy Appl. Immunol. 42, 503-515 (1972)

Gold, P., Freedman, S.O.: Demonstration of tumour-specific antigens in human colonic carcinomata by immunological tolerance and absorption techniques. J. Exp. Med. 121, 439-462 (1965)

Gordon, Y.B., Jeffrey, D., Grudzinskas, J.G., Chard, T.: Concentrations of pregnancy-specific $beta_1$-glycoprotein in maternal blood in normal pregnancy and in intrauterine growth retardation. Lancet i, 331-333 (1977)

Grudzinskas, J.G. Lenton, E.A., Gordon, Y.B., Kelso, I.M., Jeffrey, D., Sobowale, O.: Circulating levels of pregnancy-specific $beta_1$-glycoprotein in early pregnancy. Br. J. Obstet. Gynaecol. 84, 740-742 (1977)

Horne, C.H.W., Milne, G.D., Reid, I.N.: Prognostic significance of inappropriate production of PSßG by breast cancer. Lancet ii, 279-282 (1976a)

Horne, C.H.W., Towler, C.M., Pugh-Humphreys, R.G.P., Thomson, A.W., Bohn, H.: Pregnancy-specific $beta_1$-glycoprotein. A product of the syncytiotrophoblast. Experientia 32, 1197-1199 (1976b)

Horne. C.H.W., Reid, I.N., Towler, C.M., Milne. G.D.: Production of pregnancy specific $beta_1$-glycoprotein by tumours. In: 24th coll. prot. biol. fluids. PEETERS, G. (ed.), pp. 567-570. Oxford: Pergamon Press 1976

Jandial, V., Campbell, D.M., Towler, C.M., MacGillivray, I., Horne, C.H.W.: Pregnancy-specific $beta_1$-glycoprotein as indicator of placental function. Br. J. Obstet. Gynaecol. 84, 76-83 (1977)

Johnson, S.A.N., Grudzinskas, J.G., Gordon, Y.B., Al-Ani, A.T.M.: Pregnancy specific $beta_1$-glycoprotein in plasma and tissue extract in malignant teratoma of the testis. Br. Med. J. i, 951-952 (1977)

Kalashnikov, V.V., Falaleeva, D.M., Tatarinov, Y.S.: Quantitative immunoenzymatic method of evaluation of pregnancy-specific $beta_1$-glycoprotein in trophoblastic tumours. Byull. Eksp. Biol. Med. (1978) (in press)

Lin, T.S., Halbert, S.P.: Immunological comparison of various human pregnancy-associated plasma proteins. Int. Arch. Allergy Appl. Immunol. 48, 101-115 (1975)

Lin, T.S., Halbert, S.P.: Placental localization of human pregnancy-associated plasma proteins. Science 193, 1249-1252 (1976)

Lin, T.M., Halbert, S.P., Kiefer, D., Spellacy, W.N., Gall, S.: Characterization of four human pregnancy-associated plasma proteins. Am. J. Obstet. Gynecol. 118, 223-236 (1974a)

Lin, T.M., Halbert, S.P., Kiefer, D., Spellacy, W.N.: Three pregnancy-associated human plasma proteins. Purification, monospecific antisera and immunological identification. Int. Arch. Allergy Appl. Immunol. 47, 35-53 (1974b)

Lin, T.M., Halbert, S.P., Kiefer, D.: Pregnancy-associated plasma proteins during human gestation. J. Clin. Invest. 54, 576-582 (1974c)

Lin, T.M., Halbert, S.P., Spellacy, W.N.: Relation of obstetric parameters to the concentration of four pregnancy-associated plasma proteins at term in normal gestation. Am. J. Obstet. Gynecol. 125, 17-24 (1976)

McIntire, K.R., Waldmann, T.A., Moertel, C.G., Go, V.L.W.: Serum alpha-fetoprotein in patients with neoplasms of the gastrointestinal tract. Cancer Res. 35, 991-996 (1975)

Midgley, A.R., Pierce, G.B.: Immunochemical localisation of human chorionic gonadotrophin. J. Exp. Med. 115, 289-294 (1962)

Nikulina, D.M., Sokolov. A.V., Tatarinov, Y.S.: Physicochemical characteristics of specific beta$_1$-globulin from blood serum of pregnant women. Vopr. Med. Chem. 23, 324-326 (1977)

Rosen, S.W., Weintraub, R.D., Vaitukaitis, J.L., Sussmann, H.H., Herschman, J.M., Muggia, F.M.: Placental proteins and their subunits as tumour markers. Ann. Intern. Med. 82, 71-83 (1975)

Ruoslahti, E., Seppälä, M.: Studies of carcino-fetal protein. Development of a radioimmunoassay for alpha-fetoprotein. Demonstration of alpha-fetoprotein in serum of healthy human adults. Int. J. Cancer 8, 374-383 (1971)

Seppälä, M., Rutanen, E.-M., Heikinheimo, M., Jalanko, H., Engvall, E.: Detection of trophoblastic tumour activity by pregnancy-specific beta$_1$-glycoprotein. Int. J. Cancer 21, 265-267 (1978)

Shevchenko, O.P., Larina, I.M., Tatarinov, Y.S.: Immunoradioautographic study of the specificity in different species of pregnancy beta-globulins. Byull. Eksp. Biol. Med. 83, 721-723 (1977)

Tatarinov, Y.S.: Discovery of alpha-fetal globulin in sera of patients with primary cancer of the liver. Abstract, 1st int. biochem. congr. p. 274. Leningrad 1963

Tatarinov, Y.S.: Placental and embryonal proteins as antigens in immunodiagnosis of tumours. Vestn. Akad. Med. Nauk, U.S.S.R. 10, 90-93 (1977)

Tatarinov, Y.S.: A new placental protein test for the identification of trophoblastic tumours. Antibiot. Chemother. 22, 125-131 (1978)

Tatarinov, Y.S., Masyukevich, V.N.: Immunological identification of a new beta$_1$-globulin in the blood serum of pregnant women. Byull. Eksp. Biol. Med. 69, 66-68 (1970)

Tatarinov, Y.S., Sokolov, A.V.: Development of a radioimmunoassay for pregnancy-specific beta$_1$-globulin and its measurement in serum of patients with trophoblastic and nontrophoblastic tumours. Int. J. Cancer $\underline{19}$, 161-166 (1977)

Tatarinov, Y.S., Masyukevich, V.N., Emelyanchik, E.K., Kruglov, E.N., Nikulina, D.M., Parfinova, L.F.: A comparative study of alpha$_2$ and beta$_1$ globulins of the blood serum of pregnant women. Akush. Ginekol. (Mosk.) $\underline{47}$, 35-38 (1971)

Tatarinov, Y.S., Mesnyankina, N.V., Nikulina, D.M.: Immunochemical identification of 'pregnancy-zone' in blood sera of patients with hydatidiform mole and chorioepithelioma. Akush. Ginekol. (Mosk.) $\underline{50}$, 67-68 (1974a)

Tatarinov, Y.S., Mesnyankina, N.V., Nikulina, D.M., Novikova, L.A., Toloknov, B.O., Falaleeva, D.M.: Identification immunochemique de la bêta$_1$-globuline de la 'zone de grossesse' dans le sérum de malades attentes de tumeurs trophoblastiques. Int. J. Cancer $\underline{14}$, 548-554 (1974b)

Tatarinov, Y.S., Nikulina, D.A., Mesnyankina, N.V.: Immunochemical test on beta$_1$-globulin for early diagnosis of pregnancy. Lab. Delo $\underline{9}$, 547-549 (1974c)

Tatarinov, Y.S., Falaleeva, D.M., Kalashnikov, V.V.: Immunofluorescent study of beta$_1$-G-globulin of placenta in different stages of pregnancy. Akush. Ginekol. (Mosk.) $\underline{51}$, 63-65, (1975a)

Tatarinov, Y.S., Falaleeva, D.M., Elgort, D.A., Novikova, L.A., Toloknov, B.O.: Immunoradioautographic determination of beta$_1$-globulin in the blood serum of patients with trophoblastic tumours. Byull. Eksp. Biol. Med. $\underline{74}$, 86-89 (1975b)

Tatarinov, Y.S., Falaleeva, D.M., Kalashnikov, V.V.: Human pregnancy-specific beta$_1$-globulin and its relation to chorioepithelioma. Int. J. Cancer $\underline{17}$, 625-632 (1976a)

Tatarinov, Y.S., Falaleeva, D.M., Kalashnikov, V.V., Toloknov, B.O.: Immunofluorescent localization of human pregnancy-specific beta$_1$-globulin in placenta and chorioepithelioma. Nature (Lond.) $\underline{260}$, 263 (1976b.)

Tatarinov, Y.S,, Falaleeva, D.M., Kozlyaeva, G.A., Nikulina, D.M.: Human trophoblastic beta$_1$-globulin and chorioepithelioma. In: Onco-develompent and gene expression. FISHMAN, W.H., SELL, S., (eds)., pp. 463-468. New York: Academic Press 1976c

Tatarinov, Y.S., Kasatkin, Y.N., Masenko, V.P., Novikova, L.A., Toloknov, B.O., Falaleeva, D.M.: Serological test for beta$_1$-globulin 'zone pregnancy' in immunodiagnosis of trophoblastic tumours of the uterus. Vest. Acad. Med. Nauk U.S.S.R. 2, 44-48 (1976d)

Tatarinov, Y.S., Krivonosov, S.K., Petrunin, D.D., Shevchenko, O.P.,: Comparative immunochemical analysis of pregnancy-specific beta$_1$-globulin and analogous beta-globulins in mammals. Byull. Eksp. Biol. Med. $\underline{76}$, 1223-1225 (1976e)

Tatarinov, Y.S., Falaleeva, D.M., Kalashnikov, V.V., Vasiliev, M.Y.: Cellular localization of pregnancy-specific beta$_1$-globulin in human placenta. Vopr. Med. Khim. $\underline{23}$, 88-92 (1977)

Tatra, G., Breitenecker, G., Gruber, W.: Serum concentration of pregnancy-specific beta$_1$-globulin (SP-1) in normal and pathological pregnancy. Arch. Gynaekol. $\underline{217}$, 383-390 (1974)

Tatra, G., Placheta, P., Breitenecker, G.: Schwangerschaftsspezifisches beta$_1$-glykoprotein (SP-1): klinische Aspekte. Wien. Klin. Wochenschr. <u>87</u>, 279-281 (1975)

Tatra, G., Polak, S., Placheta, P.: Concentration of pregnancy-specific beta$_1$-protein SP-1 in amniotic fluid in normal and pathological pregnancies. Arch. Gynäkol. <u>221</u>, 161-165 (1976)

Towler, C.M., Jandial, V., Horne, C.H.W., Bohn, H.: A serial study of pregnancy proteins in primigravidae. Br. J. Obstet. Gynaecol. <u>83</u>, 368-374 (1976)

Towler, C.M., Horne, C.H.W., Jandial, V., Campbell, D.M., MacGillivray,I.: Plasma levels of pregnancy-specific beta$_1$-glycoprotein in complicated pregnancies. Br. J. Obstet. Gynaecol. <u>84</u>, 258-264 (1977)

Wislocki, G. B., Dempsey, E.W.: Electron microscopy of the human placenta. Anat. Rec. <u>123</u>, 133-149 (1955)

Clinical Management of Mother and Newborn

Editor: G. F. Marx

1979. 30 figures. Approx. 260 pages
ISBN 3-540-90373-9

S. N. Hassani

Ultrasound in Gynecology and Obstetrics

In collaboration with R. L. Bard

1978. 337 figures. XX, 182 pages
ISBN 3-540-90260-0

R. O. Meudt, M. Hinselmann

Ultrasonoscopic (real time) Differential Diagnosis in Obstetrics and Gynecology

Echoskopische Differential-
Diagnose in Geburtshilfe und
Gynäkologie. Sémiologie échos-
copique en obstétrique et gynéco-
logie. Semiologia ecoscópia en
obstetricia y gynecologia. Semiologia ecoscopia in ostetricia e
ginecologia

2nd revised edition 1978.
209 figures, 1 fold-out table.
X, 145 pages
ISBN 3-540-08839-3

H. Ludwig, H. Metzger

The Human Female Reproductive Tract

A Scanning Electron Microscopic
Atlas

1976. 546 microcraphs.
XI, 247 pages
ISBN 3-540-07675-1

Springer-Verlag
Berlin
Heidelberg
New York

G. Dallenbach-Hellweg

Histopathology of the Endometrium

English Translation by
F. D. Dallenbach

2nd, revised and enlarged edition
1975. 142 figures, 2 colored plates.
IX, 325 pages
ISBN 3-540-07215-2

T. Kawai

Clinical Aspects of the Plasma Proteins

1973. 278 figures, 90 tables.
20 color photos. XVI, 466 pages
ISBN 3-540-06523-7
Published by Igaku Shoin Ltd.
Tokyo. Distribution rights for
Europe (including U.K.):
Springer-Verlag

P. Stoll

Gynecological Vital Cytology

Function – Microbiology – Neoplasia. Atlas of Phase-Contrast
Microscopy

1969. 145 figures. VI, 81 pages
ISBN 3-540-04725-5
Distribution rights for Japan:
Igaku Shoin Ltd., Tokyo

H.-W. Denker

Implantation

The Role of Proteinases, and
Blockage of Implantation by
Proteinase Inhibitors. Translated
from the German.

1977. 35 figures. 123 pages
(Advances in Anatomy, Vol. 53/5)
ISBN 3-540-08479-7

Springer-Verlag
Berlin
Heidelberg
New York